NATURAL DISASTER RESEARCH, PREDICTION AND MITIGATION

ISSUES IN DISASTER RECOVERY AND ASSISTANCE

NATURAL DISASTER RESEARCH, PREDICTION AND MITIGATION

Additional books and e-books in this series can be found
on Nova's website under the Series tab.

NATURAL DISASTER RESEARCH, PREDICTION AND MITIGATION

ISSUES IN DISASTER RECOVERY AND ASSISTANCE

DONATIEN MOÏSE
EDITOR

Copyright © 2019 by Nova Science Publishers, Inc.

All rights reserved. No part of this book may be reproduced, stored in a retrieval system or transmitted in any form or by any means: electronic, electrostatic, magnetic, tape, mechanical photocopying, recording or otherwise without the written permission of the Publisher.

We have partnered with Copyright Clearance Center to make it easy for you to obtain permissions to reuse content from this publication. Simply navigate to this publication's page on Nova's website and locate the "Get Permission" button below the title description. This button is linked directly to the title's permission page on copyright.com. Alternatively, you can visit copyright.com and search by title, ISBN, or ISSN.

For further questions about using the service on copyright.com, please contact:
Copyright Clearance Center
Phone: +1-(978) 750-8400   Fax: +1-(978) 750-4470   E-mail: info@copyright.com.

NOTICE TO THE READER

The Publisher has taken reasonable care in the preparation of this book, but makes no expressed or implied warranty of any kind and assumes no responsibility for any errors or omissions. No liability is assumed for incidental or consequential damages in connection with or arising out of information contained in this book. The Publisher shall not be liable for any special, consequential, or exemplary damages resulting, in whole or in part, from the readers' use of, or reliance upon, this material. Any parts of this book based on government reports are so indicated and copyright is claimed for those parts to the extent applicable to compilations of such works.

Independent verification should be sought for any data, advice or recommendations contained in this book. In addition, no responsibility is assumed by the Publisher for any injury and/or damage to persons or property arising from any methods, products, instructions, ideas or otherwise contained in this publication.

This publication is designed to provide accurate and authoritative information with regard to the subject matter covered herein. It is sold with the clear understanding that the Publisher is not engaged in rendering legal or any other professional services. If legal or any other expert assistance is required, the services of a competent person should be sought. FROM A DECLARATION OF PARTICIPANTS JOINTLY ADOPTED BY A COMMITTEE OF THE AMERICAN BAR ASSOCIATION AND A COMMITTEE OF PUBLISHERS.

Additional color graphics may be available in the e-book version of this book.

Library of Congress Cataloging-in-Publication Data

ISBN: 978-1-53616-308-7

Published by Nova Science Publishers, Inc. † New York

CONTENTS

Preface		**vii**
Chapter 1	Disaster Recovery: Additional Actions Would Improve Data Quality and Timeliness of FEMA's Public Assistance Appeals Processing *United States Government Accountability Office*	**1**
Chapter 2	FEMA Individual Assistance Programs: In Brief *Shawn Reese*	**53**
Chapter 3	Federal Disaster Assistance: Individual Assistance Requests Often Granted, but FEMA Could Better Document Factors Considered *United States Government Accountability Office*	**63**
Chapter 4	FEMA and SBA Disaster Assistance for Individuals and Households: Application Process, Determinations, and Appeals *Bruce R. Lindsay and Shawn Reese*	**121**
Chapter 5	Disaster Assistance: FEMA Action Needed to Better Support Individuals Who Are Older or Have Disabilities *United States Government Accountability Office*	**147**

Index	**221**
Related Nova Publications	**227**

PREFACE

In both 2016 and 2017, 15 separate U.S. disasters resulted in losses exceeding $1 billion each. FEMA provides PA grants to state and local governments to help communities recover from such disasters. If applicants disagree with FEMA's decision on their PA grant application, they have two chances to appeal: a first-level appeal to be decided by the relevant FEMA regional office and, if denied, a second-level appeal to be decided within FEMA's Recovery Directorate. Chapter 1 examines the extent to which FEMA ensures the quality of its appeals data and what these data show about PA appeals inventory and timeliness; what steps FEMA has taken to improve its management of the appeals process and what challenges, if any, remain; and the extent to which FEMA developed goals and measures to assess program performance. Chapter 2 provides a short summary of the types of individual assistance programs administered by FEMA following a disaster. It also provides a summary of the criteria FEMA uses in determining which individual assistance programs may be made available to impacted areas following a major disaster declaration, and discusses a proposed rule to change these criteria. FEMA's IA program provides help to individuals to meet their immediate needs after a disaster, such as shelter and medical expenses. Chapter 3 examines the number of IA declaration requests received, declared, and denied, and IA actual obligations from calendar years 2008 through 2016, the extent to which FEMA accounts for the regulatory factors when evaluating IA requests, and any challenges FEMA regions and select states reported on the declaration process and factors and any FEMA actions to revise them. The Federal Emergency Management Agency's (FEMA's) Individual Assistance (IA) program and the Small Business Administration's (SBA's) Disaster Loan Program are the federal government's two primary sources of financial assistance to help individuals and households recover and rebuild from a major disaster as discussed in chapter 4. Hurricane survivors aged 65 and older and those with disabilities faced particular challenges evacuating to safe shelter, accessing medicine, and obtaining recovery assistance. Chapter 5 addresses challenges FEMA partners reported in providing assistance to such individuals, challenges such individuals faced accessing assistance from FEMA and actions FEMA took to address these challenges, and the extent to which FEMA has implemented its new approach to disability integration.

Chapter 1 - In both 2016 and 2017, 15 separate U.S. disasters resulted in losses exceeding $1 billion each. FEMA provides PA grants to state and local governments to help communities recover from such disasters. If applicants disagree with FEMA's decision on their PA grant application, they have two chances to appeal: a first-level appeal to be decided by the relevant FEMA regional office and, if denied, a second-level appeal to be decided within FEMA's Recovery Directorate. Each is subject to a 90- day statutory processing timeframe. GAO was asked to review FEMA's appeals process. This chapter examines: (1) the extent to which FEMA ensures the quality of its appeals data and what these data show about PA appeals inventory and timeliness; (2) what steps FEMA has taken to improve its management of the appeals process and what challenges, if any, remain; and (3) the extent to which FEMA developed goals and measures to assess program performance. GAO analyzed FEMA policies, procedures, and data on appeals and interviewed officials from headquarters and from regional offices with the highest number of pending appeals. GAO also spoke to state officials from the two states within each of the three regions with the highest number of pending appeals.

Preface ix

Chapter 2 - When the President declares a major disaster pursuant to the Robert T. Stafford Disaster Relief and Emergency Assistance Act (P.L. 93-288), the Federal Emergency Management Agency (FEMA) advises the President about types of federal assistance administered by FEMA available to disaster victims, states, localities, and tribes. The primary types of assistance provided under a major disaster declaration include funding through the Public Assistance program, Mitigation Assistance programs, and the Individual Assistance program. The Public Assistance program provides federal financial assistance to repair and rebuild damaged facilities and infrastructure. Mitigation Assistance programs provide funding for jurisdictions, states, and tribes to ensure damaged facilities and infrastructure are rebuilt and reinforced to better withstand future disaster damage. Finally, the Individual Assistance program provides funding for basic needs for individuals and households following a disaster. Eligible activities under the Individual Assistance program include funding for such things as mass care, crisis counseling, and temporary housing. FEMA advises the President on the type of individual assistance to be granted following each disaster, and works with state and local authorities in determining what assistance programs would best suit the needs within the disaster area. FEMA makes this determination based on a list of criteria designed to align federal disaster assistance with unmet needs in disaster-impacted areas. This chapter provides a short summary of the types of individual assistance programs administered by FEMA following a disaster. This chapter also provides a summary of the criteria FEMA uses in determining which individual assistance programs may be made available to impacted areas following a major disaster declaration, and discusses a proposed rule to change these criteria.

Chapter 3 - FEMA's IA program provides help to individuals to meet their immediate needs after a disaster, such as shelter and medical expenses. When a state, U.S. territory, or tribe requests IA assistance through a federal disaster declaration, FEMA evaluates the request against regulatory factors, such as concentration of damages, and provides a recommendation to the President, who makes a final declaration decision. GAO was asked to review FEMA's IA declaration process. This chapter examines (1) the number of IA declaration requests received, declared, and denied, and IA actual obligations from calendar years 2008 through 2016, (2) the extent to which FEMA accounts for the regulatory factors when evaluating IA requests, and (3) any challenges FEMA regions and select states reported on the declaration process and factors and any FEMA actions to revise them. GAO reviewed FEMA's policies, IA declaration requests and obligation data, and FEMA's RVARs from July 2012 through December 2016, the most recent years for which data were available. GAO also reviewed proposed rulemaking comments and interviewed FEMA officials from all 10 regions and 11 state emergency management offices selected based on declaration requests and other factors.

Chapter 4 - The Federal Emergency Management Agency's (FEMA's) Individual Assistance (IA) program and the Small Business Administration's (SBA's) Disaster Loan Program are the federal government's two primary sources of financial assistance to help individuals and households recover and rebuild from a major disaster. In many cases, disaster survivors find that they need assistance from both of these programs in addition to other sources of assistance including private insurance, state and local government assistance, and assistance from private voluntary organizations. Though FEMA IA and the SBA Disaster Loan Program are separate programs administered by different agencies, in many ways they are interconnected. SBA and FEMA share real-time data on disaster grant and loan approvals to identify potential duplication of benefits while providing individuals and households with federal assistance that can be used in conjunction with each other to meet recovery needs. The two programs are also interconnected in the way they are administered to determine loan and grant eligibility. Furthermore, eligibility and assistance from one source can affect eligibility and assistance from the other source. It could be argued the overlap between the two programs provides an effective means to identify duplication and provide federal assistance; however, the overlap also causes some confusion. Some in Congress are concerned that elements of the application process are not entirely known. For instance, it is unclear to some what criteria are used to determine assistance eligibility as well as how decisions are made with

Preface xi

respect to whether an applicant should be provided a grant or a loan (or both). It is also unclear whether FEMA and SBA determine eligibility on a case-by-case basis, or if eligibility criteria are applied uniformly. This chapter provides an overview of the two programs including discussions about

- how declarations put the programs into effect;
- the application process for both programs;
- the criteria used by FEMA and the SBA to determine assistance; and
- the FEMA and SBA appeal processes.

The report concludes with policy observations and considerations for Congress.

Chapter 5 - Three sequential hurricanes—Harvey, Irma, and Maria—affected more than 28 million people in 2017, according to FEMA. Hurricane survivors aged 65 and older and those with disabilities faced particular challenges evacuating to safe shelter, accessing medicine, and obtaining recovery assistance. In June 2018, FEMA began implementing a new approach to assist individuals with disabilities. GAO was asked to review disaster assistance for individuals who are older or have disabilities. This chapter addresses (1) challenges FEMA partners reported in providing assistance to such individuals, (2) challenges such individuals faced accessing assistance from FEMA and actions FEMA took to address these challenges, and (3) the extent to which FEMA has implemented its new approach to disability integration. GAO analyzed FEMA data and reviewed relevant federal laws, agency policy, and federal frameworks. GAO also interviewed state, territorial, local, and nonprofit officials in Florida, Puerto Rico, Texas, and the U.S. Virgin Islands; FEMA officials at headquarters, in regional offices, and deployed to disaster sites; and officials at relevant nonprofit organizations.

In: Issues in Disaster Recovery and Assistance   ISBN: 978-1-53616-308-7
Editor: Donatien Moïse   © 2019 Nova Science Publishers, Inc.

Chapter 1

DISASTER RECOVERY: ADDITIONAL ACTIONS WOULD IMPROVE DATA QUALITY AND TIMELINESS OF FEMA'S PUBLIC ASSISTANCE APPEALS PROCESSING[*]

United States Government Accountability Office

ABBREVIATIONS

DHS	Department of Homeland Security
FEMA	Federal Emergency Management Agency
GPRA	Government Performance and Results Act of 1993
GPRAMA	GPRA Modernization Act of 2010
OIG	Office of the Inspector General
PA	Public Assistance
PAAB	Public Assistance Appeals and Audits Branch

---

[*] This is an edited, reformatted and augmented version of United States Government Accountability Office; Report to Congressional Requesters, Publication No. GAO-18-143, dated December 2017.

RFI	Request for Information
SOP	Standard Operating Procedures
Stafford Act	Robert T. Stafford Disaster Relief and Emergency Assistance Act

WHY GAO DID THIS STUDY

In both 2016 and 2017, 15 separate U.S. disasters resulted in losses exceeding $1 billion each. FEMA provides PA grants to state and local governments to help communities recover from such disasters. If applicants disagree with FEMA's decision on their PA grant application, they have two chances to appeal: a first-level appeal to be decided by the relevant FEMA regional office and, if denied, a second-level appeal to be decided within FEMA's Recovery Directorate. Each is subject to a 90- day statutory processing timeframe.

GAO was asked to review FEMA's appeals process. This chapter examines: (1) the extent to which FEMA ensures the quality of its appeals data and what these data show about PA appeals inventory and timeliness; (2) what steps FEMA has taken to improve its management of the appeals process and what challenges, if any, remain; and (3) the extent to which FEMA developed goals and measures to assess program performance. GAO analyzed FEMA policies, procedures, and data on appeals and interviewed officials from headquarters and from regional offices with the highest number of pending appeals. GAO also spoke to state officials from the two states within each of the three regions with the highest number of pending appeals.

WHAT GAO RECOMMENDS

GAO is making four recommendations, including that FEMA implement a consistent approach for tracking appeals and ensuring data

Disaster Recovery 3

quality, develop a workforce plan, and develop measurable goals for processing first-level appeals. FEMA concurred with all four recommendations.

WHAT GAO FOUND

Weaknesses in the quality of Federal Emergency Management Agency's (FEMA) Public Assistance (PA) appeals data limit its ability to oversee the appeals process. For example, FEMA's data are inaccurate and incomplete because regional offices do not consistently track first-level appeals and FEMA does not have processes to ensure data quality. When GAO discussed these weaknesses with FEMA officials, they acknowledged them and provided GAO with corrected data for January 2014 through July 2017. GAO's analyses of the corrected data show fluctuations in the appeal inventory from year to year depending on the number of disasters declared and delays in processing. For example, as shown in the figure, only 9 percent of first-level and 11 percent of second-level appeals were processed within the 90-day statutory timeframe.

FEMA has taken steps to improve its management of the appeals process— including issues that GAO and the Department of Homeland Security's Office of Inspector General identified in 2008 and 2011. For example, FEMA increased its appeal staffing levels and developed standard operating procedures. Despite these efforts, FEMA continued to face a number of workforce challenges that contributed to processing delays, such as staff vacancies, staff turnover, and delays in training. FEMA has not developed a workforce staffing plan to identify hiring, training, and retention needs across its headquarters and regional offices, though FEMA officials acknowledge the potential benefits of having such a plan and stated that they are focused on filling vacancies. In the absence of a workforce plan, FEMA will continue to experience workforce challenges that could further contribute to delays in processing appeals.

FEMA has not established goals and measures for assessing first-level appeals processing performance, but has done so for second-level appeals.

FEMA views establishing these first-level goals and measures as the responsibility of its regional offices. Without goals and measures, FEMA is limited in its ability to assess the efficiency and effectiveness of its overall appeals process and identify and address weaknesses that may lead to delays in making appeal decisions.

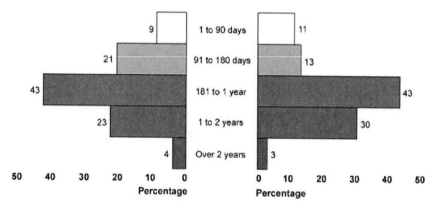

Source: GAO analysis of Federal Emergency Management Agency (FEMA) data. | GAO-18-143.

Processing Times for Decided Appeals, Based on Appeals FEMA Received between January 2014 and July 2017.

December 15, 2017
Congressional Requesters

In both 2016 and 2017, 15 separate U.S. disasters resulted in losses exceeding $1 billion each, and 2017 has tied the record pace for frequency of billion-dollar disasters occurring in the calendar year. For the communities that have been affected by these disasters, critical aspects of response and recovery, such as removing debris, and rebuilding the infrastructure of state and local schools, roads, and utilities, may take years. The damage from such disasters points to the strains on state and local resources in both response and recovery, especially in the event of back-to-back catastrophic disasters, such as those witnessed during the 2017 hurricane season.

Disaster Recovery 5

Each year, the federal government obligates billions of dollars through programs and activities that provide assistance to state, local, territorial, and tribal governments; individuals; and certain nonprofit organizations that have suffered injury or damages from major disasters or emergency incidents, such as hurricanes, tornados, and fires.[1] We recently reported that, from fiscal years 2005 through 2014, the federal government obligated at least $277.6 billion in disaster assistance through a range of programs.[2] One such program, Public Assistance (PA), provides grants to state, local, territorial, and tribal governments and certain nonprofit organizations following a disaster. Administered through the Federal Emergency Management Agency (FEMA), a component of the Department of Homeland Security (DHS), PA provides financial assistance for debris removal; emergency protective measures; and the repair, replacement, or restoration of disaster-damaged, publicly owned facilities, and the facilities of certain private nonprofit organizations.[3] From fiscal years 2009 through 2016, FEMA obligated more than $36 billion in grants for such projects.

As in the case of many federal grant programs, FEMA provides PA funds to a state recipient which, in turn, passes these funds along to a local entity, based on an application for assistance.[4] Applicants who request disaster assistance through the PA program are entitled to appeal any decision regarding how FEMA determined their eligibility for PA grant funds, including obligated amounts of PA funding.[5] FEMA can also

---

[1] An obligation is a definite commitment that creates a legal liability of the government for the payment of goods and services ordered or received, or a legal duty on the part of the United States that could mature into a legal liability by virtue of actions on the part of the other party beyond the control of the United States. Payment may be made immediately or in the future. An agency incurs an obligation, for example, when it places an order, signs a contract, awards a grant, purchases a service, or takes other actions that require the government to make payments to the public or from one government account to another.

[2] GAO, *Federal Disaster Assistance: Federal Departments and Agencies Obligated at Least $277.6 Billion during Fiscal Years 2005 through 2014*, GAO-16-797 (Washington, D.C.: Sept. 22, 2016).

[3] 42 U.S.C. § 5172.

[4] For purposes of this report, we refer to the recipient as the state. However, territories and tribal governments are also considered recipients. We also refer to the applicant as the entity that is appealing a PA decision.

[5] 42 U.S.C. § 5189a.

6 *United States Government Accountability Office*

deobligate PA funding when it finds, for example, that ineligible work was performed, incurred costs were later deemed unreasonable, or improper procurement methods were used.[6] Applicants can also appeal decisions resulting from FEMA audits of PA projects, even after a project has been completed and closed as long as the appeal is filed in a timely manner.

To appeal a decision, applicants are afforded two opportunities: (1) a first-level appeal to the relevant FEMA regional office to be decided by the Regional Administrator and (2) a second-level appeal to be decided at FEMA headquarters by the Assistant Administrator for the Recovery Directorate or the PA Division Director through a delegation of authority. FEMA regulations implement the statutory time frames for applicants to file an appeal, for the state to transmit the appeal to FEMA, and for FEMA to respond to the appeal.[7]

We and the DHS Office of the Inspector General (OIG) have identified a number of issues related to FEMA's PA program, including its PA appeals process. For example, in 2008, as part of our review of FEMA's administration of the PA program, we found that, following Hurricanes Katrina and Rita, FEMA often did not make decisions on applicant appeals within statutorily required time frames.[8] In 2011, the DHS OIG also identified areas for improvement in the PA appeals process, including the timeliness of processing an appeal and communicating with applicants about the status of their appeals.[9] In 2013, FEMA stood up the Public Assistance Appeals Branch (PAAB) within the Office of the Recovery Directorate in an effort to respond to these concerns, adding an auditing component to the Branch in 2014.

You asked us to review aspects of FEMA's management of the PA appeals process. This review examines: (1) the extent to which FEMA ensures quality in its data on appeals and what FEMA data show about its

---

[6] Funds can also be deobligated when an agency cancels or downwardly adjusts previously incurred obligations.

[7] 44 C.F.R. § 206.206.

[8] GAO, *Disaster Recovery: FEMA's Public Assistance Grant Program Experienced Challenges with Gulf Coast Rebuilding*, GAO-09-129 (Washington, D.C.: Dec. 18, 2008).

[9] Department of Homeland Security, Office of Inspector General, Opportunities to Improve FEMA's Public Assistance Appeals Process, DHS OIG-11-49 (Washington, D.C.: March 2011).

appeals inventory and timeliness for appeals decisions; (2) what steps FEMA has taken to improve its management of the appeals process and what challenges, if any, remain; and (3) the extent to which FEMA has developed goals and measures to assess the PA appeal program's performance.

To address the first objective, we obtained and analyzed data from FEMA on all first- and second-level appeals that the agency received between January 2014 and July 2017. We focused on this time frame because it contained the most complete and available data at the time of our review. We identified various discrepancies in the first-level appeal data, which we discussed with knowledgeable FEMA staff and present later in this chapter. In response to our discussions, FEMA provided us with corrected data to address the identified discrepancies. After obtaining the corrected data and making adjustments to our analysis based on our discussions with FEMA officials, we determined that the appeals data from FEMA were sufficiently reliable to provide information on PA appeals, including appeals inventories, outcomes, amounts in dispute, and processing times that we present in this chapter. We also obtained and analyzed FEMA policies and procedures related to tracking appeals data, including those related to regional offices, and evaluated them using *Standards for Internal Control in the Federal Government*.[10]

To address the first and second objectives, we also administered semistructured interviews to officials from 3 of FEMA's 10 regional offices (Regions II, IV, and VI) with the highest number of first- and second-level pending appeals. We asked these officials about their efforts to process and track appeals, what improvements had been made regarding how PA appeals are processed, as well as any challenges they have faced in processing PA appeals since 2013.To select these offices, we obtained data from FEMA on first- and second-level appeals that were pending a decision, as of October 31, 2016. Collectively, appeals from these 3 regional offices represented 69 percent of all pending first-and second-level appeals FEMA had received as of October 31, 2016. We focused on

---

[10] GAO, *Standards for Internal Control in the Federal Government*, GAO-14-704G (Washington, D.C.: September 2014).

this time frame because it contained the most recent data for our selection of FEMA regional offices at the time we began our review. To obtain additional perspective on what, if any, challenges remain in FEMA's management of the appeals process, we also interviewed state emergency management officials in six states (two states in each of the corresponding 3 FEMA regional offices.) (See Figure 1).

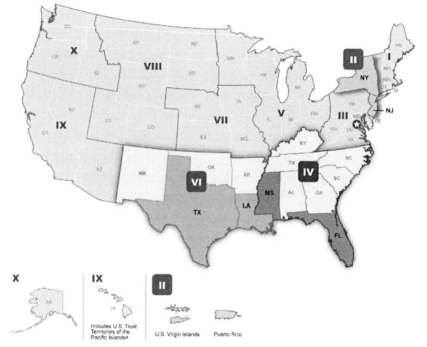

Source: GAO analysis based on Federal Emergency Management Agency (FEMA) information (data); Map Resources (map). | GAO-18-143.

Figure 1. FEMA Regional Offices and State Emergency Management Offices Included in GAO's Review.

To additionally address the second objective, we reviewed FEMA documentation, such as standard operating procedures (SOP) and policies, directives, internal staffing requests, appeals analyst position descriptions, and other internal memoranda. We used these sources to identify what steps FEMA had taken to improve its management of the appeals process since 2013. We also used this information to supplement our understanding of the challenges FEMA, including its regional officials, raised during our interviews discussed above.

To address the third objective, we obtained FEMA internal reports—including briefs and newsletters—and performance plans to identify goals and measures FEMA had developed to assess the appeals program. We assessed that information against federal internal control standards[11] and leading practices we have identified in our prior work for managing for results to determine the extent to which FEMA had developed goals and measures to assess program performance.[12]

For all three objectives, we reviewed relevant legislation and interviewed officials in PAAB and FEMA's Recovery Directorate. See appendix I for a more detailed description of our scope and methodology.

We conducted this performance audit from July 2016 to December 2017 in accordance with generally accepted government auditing standards. Those standards require that we plan and perform the audit to obtain sufficient, appropriate evidence to provide a reasonable basis for our findings and conclusions based on our audit objectives. We believe that the evidence obtained provides a reasonable basis for our findings and conclusions based on our audit objectives.

BACKGROUND

FEMA's Public Assistance Process

The Robert T. Stafford Disaster Relief and Emergency Assistance Act (Stafford Act), as amended, defines FEMA's role during disaster response and recovery.[13] One of the principal programs that FEMA operates to

---

[11] GAO-14-704G.

[12] GAO, *Managing for Results: Agencies' Trends in the Use of Performance Information to Make Decisions*, GAO-14-747 (Washington, D.C.: Sept. 26, 2014) and *Executive Guide: Effectively Implementing the Government Performance and Results Act*, GAO/GGD-96-118 (Washington, D.C.: June 1, 1996).

[13] 42 U.S.C. § 5121 et seq. Under the Stafford Act, the governor of a state may request a declaration of a major disaster when effective response and recovery are beyond the

10   *United States Government Accountability Office*

fulfill its role is the PA program. PA is a complex and multistep grant program administered through a partnership between FEMA and states, which pass these funds along to eligible local grant applicants. Thus, PA entails an extensive paperwork and review process between FEMA and the state based on specific eligibility rules that outline the types of damage that can be reimbursed by the federal government and steps that federal, state, local, territorial, and tribal governments as well as certain nonprofit organizations must take in order to document eligibility. The complexity of the process led FEMA to re-engineer the PA program, which FEMA has referred to as its "new delivery model." FEMA began testing the new delivery model at select disaster locations in 2015, in preparation for implementing it nationwide for all new disasters.[14] On September 12, 2017, FEMA announced that the new delivery model would be used in all future disasters unless determined infeasible in a particular instance.[15]

The process begins after FEMA determines that the applicant meets eligibility requirements. FEMA then works with the state and the applicant to develop a project worksheet describing the scope of work and estimated cost.[16] Once FEMA and the applicant agree on the damage assessment, scope of work, and estimated costs, the PA grant obligation is determined. After FEMA approves a project, funds are obligated—that is, they are

---

     capabilities of the state and affected local governments and that federal assistance is necessary. 42 U.S.C. § 5170.

[14] As part of this effort, FEMA redesigned processes for developing, reviewing, and approving PA grant applications. The agency is also developing new PA staff positions, implementing a centralized and standardized grant processing approach, and developing a new information system to better maintain and share grant documentation. FEMA officials also reported taking steps to better incorporate hazard mitigation during the PA grant process. In November 2017, we completed a review of FEMA's progress in implementing the new delivery model. See GAO, *Disaster Assistance: Opportunities to Enhance Implementation of the Redesigned Public Assistance Grant Program*, GAO-18-30 (Washington, D.C.: Nov. 8, 2017).

[15] At the time of our review, FEMA planned to implement the new delivery model for all future disasters beginning in January 2018. However, according to FEMA officials, historic disaster activity during the hurricane season in 2017 accelerated this timeline as the states of Texas and Florida requested use of the new delivery model in response to hurricanes Harvey and Irma and FEMA agreed to allow the states' use of the new delivery model. FEMA's decision was made, in part, because of the resources that would be necessary to support the simultaneous delivery of two PA program models requiring distinct positions, processes, training, and tools over an extended period of time.

[16] Unlike a typical federal grant program, there are no caps on the amount of funding an applicant can receive under the PA program as long as the project meets eligibility requirements.

Disaster Recovery

made available—to the state recipient, which, in turn, passes the funds along to applicants.[17]

Applicants may appeal project decisions if they disagree with FEMA's decisions on project eligibility, scope of damage, or cost estimates. Appealable decisions can occur at various times during the PA grant process, including during project closeout as long as they meet applicable time limits.[18]

FEMA's PA Appeals Process

Figure 2 summarizes the first- and second-level appeals process under FEMA's PA program. The first-level appeals process begins after FEMA makes its determination on a project for PA grant assistance. Within 60 days of receiving FEMA's initial determination, the applicant must file an appeal through the state to the relevant FEMA regional office.[19]

The state must forward the appeal and a written recommendation to the relevant FEMA regional office within 60 days. In reviewing the first-level appeal before forwarding it to FEMA, the state has discretion to support or oppose all or part of the applicant's position in the appeal.

Under the Stafford Act, the FEMA regional office shall render a decision within 90 days from the date it received the first-level appeal from the state.[20] The PA appeals process can take longer if regional officials issue a request for information (RFI) to the applicant or request technical advice from subject-matter experts. According to a senior PAAB official, a regional office may issue an RFI or seek technical advice when an applicant's appeal is incomplete, lacks referenced documentation, or raises

---

[17] For purposes of this report, and as noted earlier, we refer to the recipient as the state. However, territories and tribal governments are also considered recipients. We also refer to the applicant as the entity that is appealing a PA decision.

[18] To be timely, an appeal must be filed by an applicant within 60 days of receiving notice of the action that is being appealed.

[19] The appeal must be made in writing and contain a documented justification supporting the applicant's position, specifying the monetary figure in dispute and the provisions in federal law, regulation, or policy with which the applicant believes the initial action was inconsistent.

[20] 42 U.S.C. § 5189a.

additional eligibility concerns. The regional office may issue multiple RFIs prior to rendering a final decision on an appeal. Within 90 days following the receipt of the requested additional information or following expiration of the period for providing the information, FEMA is to notify the state in writing of the disposition of the appeal.

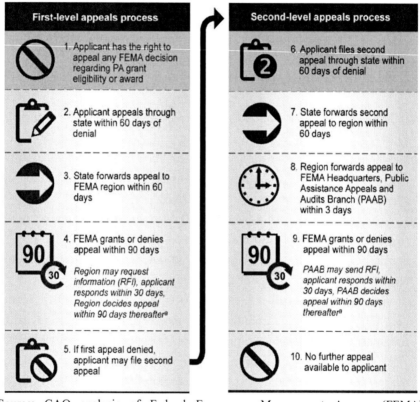

Source: GAO analysis of Federal Emergency Management Agency (FEMA) information. | GAO-18-143.

[a]FEMA may issue an RFI to an applicant multiple times before it renders a decision on an appeal.

Figure 2. Summary of FEMA's First- and Second-Level Public Assistance Appeals Process.

Disaster Recovery 13

Regional Administrators are responsible for authorizing a final decision on a first-level appeal.[21] A decision may result in an appeal being granted, partially granted, or denied. An appeal is considered granted when FEMA has approved the relief requested by the applicant as part of the appeal. An appeal is considered partially granted when FEMA has approved a portion of the relief requested by the applicant. An appeal is considered denied when FEMA has decided not to approve the relief requested by the applicant. If the regional office is considering denying or partially granting a first appeal, it must issue an RFI to provide applicants with a final opportunity to supplement the administrative record (i.e., the documents and materials considered in processing a first-level appeal), which closes upon issuing a first-level appeal decision.[22] According to a senior PAAB official, this process adds additional time to first-level appeal processing, but ensures that FEMA has considered all relevant and applicable documentation. The applicant may file a second-level appeal through the state within 60 days of receiving a first-level appeal decision. The second-level appeal must explain why the applicant believes the original determination is inconsistent with law or policy and the monetary amount in dispute. The state then has 60 days to provide a written recommendation to FEMA. In reviewing the second-level appeal, just as with the first-level appeal, the state has discretion to support or oppose all or part of the applicant's position in the appeal.

The FEMA Assistant Administrator for Recovery or the PA Division Director through a delegation of authority shall render a decision within 90 days of receipt of the second-level appeal from the state. All second-level appeal decisions are posted to FEMA's website, so applicants can review the previous decisions. As is the case with first-level appeals, the PA appeals process can take longer if PAAB officials request additional information or technical advice on an appeal. These requests must also

---

[21] Regional Administrators in each of the regional offices are responsible for the day-to-day management and administration of regional activities and staff, and report directly to the FEMA Administrator.

[22] This record may include, for example, supporting backup documentation, correspondence, photographs and technical reports, materials submitted by the applicant, and other relevant information.

14 *United States Government Accountability Office*

include a date by which the information must be provided. According to a senior PAAB official, RFIs are seldom issued for second-level appeals because the administrative record is closed after a decision is rendered on a first-level appeal. Similarly, this official told us that technical advice is rarely sought for second-level appeals because such issues are typically explored during the first-level appeal process. Within 90 days following the receipt of the requested additional information or following expiration of the period for providing the information, FEMA is to notify the state in writing of the disposition of the appeal. FEMA's response to a second-level appeal is the last and final agency decision in the appeals process.

Organization of FEMA's PA Appeals Program

Located within the Recovery Directorate, PAAB maintains overall responsibility for administering and overseeing FEMA's PA appeals program. Among other things, PAAB is responsible for ensuring that all appeal decisions are issued within regulatory timelines by developing and maintaining SOPs; arranging for supplemental staff support as needed; providing regular updates for both first- and second-level appeal decisions through a range of communications; and providing training to certify PA program staff on appeals processing.

PA program appeals staff in each of FEMA's 10 regional offices are responsible for processing first-level appeals, while PAAB staff in FEMA's Headquarters office are responsible for processing second-level appeals. Accordingly, each regional office is required to follow FEMA's Directive, Manual, and Regional SOP for processing first-level appeals, consistent with those established for second-level appeals. FEMA regional offices are also required to forward all incoming second-level appeals to PAAB. In addition, regional office staff must, within 3 business days of receiving a first-level appeal from a state, provide an electronic copy of the appeal to the PAAB via FEMA's shared workspace SharePoint site.[23] As

---

[23] SharePoint is a Microsoft web-based tool designed to store, organize, share, and access information.

Disaster Recovery   15

noted in FEMA's Recovery Directorate Appeals Manual, this step enables PAAB staff to identify and track appeals issues and trends in development across all FEMA regions. The roles and responsibilities for both first-and second-level appeals are defined in FEMA's SOPs. For example, certified appeals analysts are responsible for reviewing incoming appeals for completeness, researching and drafting appeal decisions, and generating RFIs. Lead appeals analysts are the first-line reviewers of appeal decisions and RFIs, and provide guidance on PA program and policy issues, coordinate appeals assignments, and review work of appeals analysts. Further, appeals coordinators are responsible for receiving incoming appeals, tracking the processing of those appeals, updating the appeal status, and processing other appeals-related correspondence and reports.

Prior Reviews Examining the PA Appeals Process

We have identified a number of issues related to FEMA's management of the PA appeals program in our prior audit work, as has DHS's OIG. In our 2008 review of FEMA's administration of the PA program following Gulf Coast Hurricanes Katrina and Rita, we identified challenges related to applicants' experience with appeal processing delays and that FEMA often did not make decisions on appeals within the 90-day statutory time frame.[24] Other challenges identified were that FEMA did not inform some applicants of the status of their appeal, or, in some cases, assure them of the independence of the FEMA officials making appeal decisions. Specifically, some applicants perceived there to be a conflict of interest because the PA program staff responsible for reviewing appeals was the same staff that had made the PA project decision that was being appealed. We did not make recommendations to FEMA to address these challenges in our 2008 review, but rather described the challenges as part of the status of overall Gulf Coast hurricane recovery efforts. In 2011, DHS's OIG conducted a review of FEMA's PA appeals process and made a number of

---

[24] GAO-09-129.

recommendations aimed at improving aspects of the process, including the timeliness of appeals processing, appeals staffing, and the accuracy of appeals data. As in our 2008 review, the OIG identified appeal processing delays occurring at both FEMA regional offices and at headquarters. For example, the report found that appeals were left open for long periods of time and that some regional offices as well as FEMA headquarters took more than 90 days to issue a decision on first- and second-level appeals. Further, the OIG review found that staffing approaches employed by individual regional offices contributed to processing delays and varying processing timeframes. For example, the management and processing of first-level appeals varied by FEMA regional office in that some regional offices assigned staff specifically to review appeals, while other offices assigned staff to appeals processing as part of their other responsibilities within the PA Program, such as determining eligibility for PA assistance. Further, second-level appeals were processed by various offices within FEMA headquarters, and FEMA had not established guidelines to complete work within a specific timeframe. Moreover, the OIG review found inaccuracies with FEMA's system for tracking appeal processing times for second-level appeals, resulting in unreliable information being reported to FEMA management regarding compliance with the 90-day statutory time frame. Lastly, the OIG reported that some applicants had been unable to obtain information on the status of their appeals and that FEMA did not provide meaningful feedback to resolve applicants' inquiries.

WEAKNESSES EXIST IN FEMA'S OVERSIGHT OF DATA QUALITY, BUT CORRECTED FEMA DATA SHOWED FLUCTUATIONS IN APPEAL INVENTORY AND DELAYS IN PROCESSING

Our review of FEMA data that track first- and second-level appeals showed weaknesses in the agency's data quality practices that affect program oversight. For example, we found that FEMA regional offices do not track first-level appeals data consistently or update this data regularly, resulting in missing data entries. Further, we found that FEMA's appeal tracking process does not ensure data quality, limiting FEMA's ability to use the data for making decisions on and improvements to the PA appeals process. During our review, we discussed with FEMA officials the discrepancies we found with these appeals data. FEMA officials acknowledged these data quality issues and provided us with corrected data to address these discrepancies for our analysis in this chapter. Our analysis of the corrected FEMA data showed that, between January 2014 and July 2017, FEMA received over 1,400 first- and second-level appeals with amounts in dispute totaling about $1.5 billion. Across all years, first-level appeals accounted for the majority of appeals, though the number of appeals fluctuated widely each year. Over the same period, only a small percentage of first-and second-level appeals were processed within the 90-day statutory time frame.

Weaknesses in FEMA's Tracking and Data Quality Practices Affect Program Oversight

To administer and oversee the PA appeals program, FEMA collects and tracks information on first- and second-level appeals. Based on FEMA's SOP, the agency uses this information to identify trends throughout the appeals process and identify areas in need of improvement. Specifically, PAAB uses two Excel spreadsheets for collecting and analyzing first- and second-level appeals data. The spreadsheet for collecting second-level appeals data is updated and maintained by PAAB, while the spreadsheet for first-level appeals is based on input from FEMA's 10 regional offices.

Based on our detailed review of the spreadsheets, they contain numerous data fields on the status and outcomes of first-level appeals, such as the date the regional office received the appeal, the date an RFI was issued, the date the Regional Administrator signed the decision, the

18   *United States Government Accountability Office*

amounts being disputed by the applicant, and keyword information regarding the subject of the appeal.

PAAB requests that regional offices update appeal information in the first-level appeal spreadsheet as changes occur on an appeal. PAAB then uses this data to assess trends in regional office appeals processing, which it includes in various performance and other internal reports that are shared with FEMA management and used to monitor the program. According to PAAB officials, such information provides valuable support to PAAB as well as the PA program by sharing information about filings, progress, and PA program decision making. However, while PAAB's tracking efforts help maintain visibility over and provide some monitoring of the appeals processing, we found that data fields for first-level appeals were not consistently reported or updated and that PAAB has no processes to ensure the quality of these data. As a result, data on first-level appeals may not have the accuracy needed for effective reporting and oversight efforts.

FEMA Regional Offices Do Not Track Appeals Information Consistently or Update First-Level Appeal Information Regularly

Our review of first-level appeals data showed that, between January 2014 and July 2017, regional offices did not consistently report first-level appeal information for a number of the key data fields in the PAAB first-level appeal tracking spreadsheet. Specifically, we found missing entries for the majority of the spreadsheet's 50 data fields. For example, we found that about one-third of the time, regional offices had not completed the data field for amounts being disputed by the applicant for pending appeals or indicated whether or not money was in dispute in the appeal.[25] We also

---

[25] As we discussed earlier in this report and explain in greater detail in appendix I, we identified various discrepancies in the data, which we discussed with knowledgeable FEMA staff. In response to our discussions, FEMA provided us with corrected data to address the identified discrepancies. After obtaining the corrected data and making adjustments to our analysis, we determined the appeals data from FEMA were sufficiently reliable to provide information on Public Assistance appeals that we present in this report.

found that the regional offices had generally not entered the date that the regional appeal staff had completed an initial review of the appeal—99 percent of entries were missing for this field. In another example, the data field that captures keywords was missing in over 33 percent of data entries. PAAB officials told us that keywords are an important tool for understanding the root causes of an appeal.

Further, we found a number of missing data entries for key dates for one regional office in particular. Specifically, this office had not recorded entries for any of the data fields related to key dates in the appeal process, such as the date the first-level appeal was assigned to an appeals analyst, the date the appeal was reviewed by the Regional Administrator, and the date the first-level appeal decision was sent to and received by the applicant. PAAB officials told us that PAAB uses these dates to calculate appeal processing times as part of its effort to evaluate trends in appeal information and identify potential areas for improvement, including timeliness. However, officials from this regional office told us the office does not consistently update information in the PAAB first-level appeal tracking spreadsheet and does not consider it a priority. Rather, the office considers the actual processing of first-level appeals a priority.

In addition, our analysis of first-level appeals data also showed that there was limited standardization of recording entries within fields. For example, officials in one of the three regional offices in our review told us that, in some instances, they combine first-level appeals that involve direct administrative costs and record them as a single appeal. However, the other two regional offices in our review told us they do not combine individual appeals that involve direct administrative costs. Rather, they count each as a separate appeal. The lack of standardization in the way appeals are counted could result in some types of appeals being over- or under-reported. More specifically, these inconsistencies may affect PAAB's ability to compare appeal processing capacity between regional offices and accurately report the regions' performance.

FEMA's Appeal Tracking Process Does Not Ensure Data Quality

PAAB officials acknowledged inconsistencies in first-level appeals reporting, but noted that under FEMA's SOP, the regional offices are responsible for entering first-level appeal information. According to PAAB officials, this responsibility is emphasized during training sessions with appeal staff. However, we found that FEMA has no automated data entry checks for information the regions enter into PAAB's first-level appeal tracking spreadsheet and does not monitor data fields for missing or conflicting data. Regional offices do not have a means for electronically uploading first-level appeal information to PAAB and must manually input data into the spreadsheet. PAAB's process then simply confirms receipt of the information through an email exchange with the regional office staff who manually input the information.

PAAB officials told us that they rely on regional office appeal staff to confirm and validate the first-level appeals data that are provided to PAAB for internal reporting. However, PAAB has no independent and consistent method of verifying the accuracy of the appeals data reported to it by the regional offices. PAAB officials also noted that there is no systematic process or method to identify these errors and generate an error report.

Moreover, another limitation that we identified in the spreadsheet used by the regional offices is that it is not clear what blank data fields represent— that is, whether data does not exist or whether data that exists were not recorded. PAAB officials acknowledged that blank data fields in the first-level appeal tracking spreadsheet created reporting challenges, such as whether the data field was not applicable to a particular appeal, the appeal staff for a particular region did not collect this information, or existing information was not recorded. We also identified a number of other data entries that were erroneously recorded as first-level appeals. Specifically, the information entered related to requests for adjustments to PA project funding and should not have been entered into the tracking spreadsheets as appeals.

Standards for Internal Control in the Federal Government advises management to process data into quality information that is appropriate, current, complete, accurate, accessible, and provided on a timely basis. Additionally, management should evaluate processed information, make revisions when necessary so that the information is quality information, and use the information to make informed decisions.[26] By developing and implementing processes and procedures to ensure a uniform and consistent approach for tracking first-level appeals data and better integrating regional trackers with PAAB's own first-level appeals tracker, PAAB will have greater assurance that it is collecting the comprehensive and complete appeals processing performance information it needs from the regional offices. Further, by identifying data discrepancies and other anomalies in its data queries and the resulting datasets, PAAB may be able to identify overall weaknesses in its data recording process, thereby allowing it to more accurately report on first-level appeals information. Without obtaining quality appeals data, FEMA will not be able to identify existing gaps in its appeals information and address areas in need of improvement, such as meeting statutory timeframes.

Corrected FEMA Data Showed Fluctuations in Appeal Inventory

After we shared our concerns about the appeals data with FEMA officials, they corrected the errors in their data and provided us a corrected data set to use for our analysis in this chapter. Based on our analysis of this corrected data we determined that, from January 2014 to July 2017, FEMA received over 1,445 first- and second-level appeals with amounts in dispute totaling about $1.5 billion.[27]

---

[26] GAO-14-704G.

[27] The time period for the appeals data discussed in this report includes: first-level appeals that FEMA received between January 1, 2014, and July 12, 2017, and second-level appeals that FEMA received between January 1, 2014, and July 6, 2017. The July 12, 2017, and July 6, 2017, dates are also the end of our period of analysis for the first- and second-level appeals data (respectively) in our review. For example, our analysis of the number of first-level

22 *United States Government Accountability Office*

Across all years, first-level appeals accounted for the majority of appeals, though the number of appeals fluctuated widely between years. (See Figure 3.) FEMA officials told us that the number of appeals they received has varied year to year and that increases or decreases in appeals are largely a function of the number of and severity of disaster events. That is, the greater the number of disasters declared and the more extensive the damage, the greater the number of PA program grants FEMA may issue to applicants, which in turn, may affect the likelihood that an applicant will appeal a FEMA decision regarding a grant.

FEMA issued a decision on 953 of the appeals it received between January 2014 and July 2017. As shown in Table 1, another 349 appeals were pending and awaiting a decision as of July 2017. The remaining 143 appeals were withdrawn by the applicant during the appeals process.

Table 1. Disposition of appeals FEMA received between January 2014 and July 2017

Appeals	Total pending	Total decided	Total withdrawn	Total
First-level	287	779	137	1203
Second-level	62	174	6	242
Total	349	953	143	1,445

Source: GAO analysis of Federal Emergency Management Agency (FEMA) data. I GAO-18-143.

Note: FEMA data for 2017 includes only first-level appeals that FEMA received between January 1, 2017, and July 12, 2017, and second-level appeals that FEMA received between January 1, 2017, and July 6, 2017. The July 12, 2017, and July 6, 2017, dates are also the end of our period of analysis for the first- and second-level appeals data (respectively) in our review.

Our analysis of the corrected FEMA data also found that, for appeals received between January 2014 and July 2017, total first- and second-level pending and decided appeals involved amounts in dispute totaling over

---

appeals FEMA received between January 2014 and July 2017 reflects an "as of date" of July 12, 2017. For a more detailed discussion of this data, see appendix I of this report. From the total number of appeals received, we excluded four second-level appeals that had been remanded or rescinded. According to FEMA, a rescinded appeal occurs when a regional office has made a procedural error during the first-level appeal process (e.g., failing to properly close the administrative record) and the Regional Administrator opts to correct the matter by reopening the first-level appeal process.

$1.3 billion (excluding the 143 appeals that were withdrawn by the applicant during the appeals process).[28]

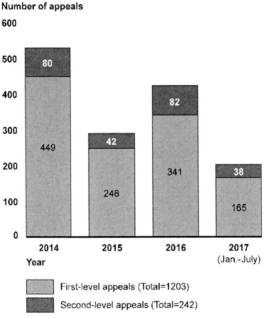

Source: GAO analysis of Federal Emergency Management Agency (FEMA) data. | GAO-18-143.

Note: FEMA data for 2017 includes only first-level appeals that FEMA received between January 1, 2017, and July 12, 2017, and second-level appeals that FEMA received between January 1, 2017, and July 6, 2017. The July 12, 2017, and July 6, 2017, dates are also the end of our period of analysis for the first- and second-level appeals data (respectively) in our review.

Figure 3. Number of Appeals FEMA Received Between January 2014 and July 2017.

As shown in Figure 4, at least a third of both first-and second-level pending and decided appeals (35 percent and 44 percent, respectively) involved amounts in dispute that ranged from $1 to $99,999. Less than 10 percent of both first- and second-level pending and decided appeals (9

---

[28] Amounts in dispute represent the total amount of PA grant funds an applicant has requested as part of an appeal. This amount, however, may change throughout the appeals process. In addition, not all appeals include an amount in dispute. For example, some appeals may involve PA program eligibility or timing requirements that do not include a monetary value.

percent and 8 percent, respectively) did not involve monetary amounts in dispute.

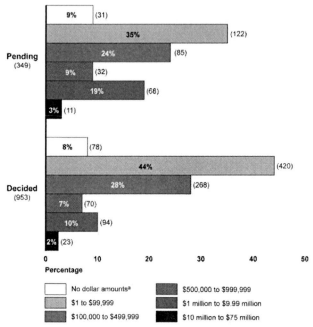

Source: GAO analysis of Federal Emergency Management Agency (FEMA) data. | GAO-18-143.

Note: FEMA data for 2017 includes only first-level appeals that FEMA received between January 1, 2017, and July 12, 2017, and second-level appeals that FEMA received between January 1, 2017, and July 6, 2017. The July 12, 2017, and July 6, 2017, dates are also the end of our period of analysis for the first- and second-level appeals data (respectively) in our review. In addition, amounts in dispute represent the total amount of PA grant funds an applicant has requested as part of an appeal. This amount, however, may change throughout the appeals process. In addition, not all appeals include an amount in dispute. For example, some appeals may involve PA program eligibility or timing requirements that do not include a monetary value. Some percentages do not add to 100 due to rounding.

[a]No dollar amounts represent those appeals that did not include a monetary value, such as those involving PA program eligibility or timing requirements.

Figure 4. Amounts in Dispute for First- and Second-Level Appeals FEMA Received between January 2014 and July 2017.

Disaster Recovery

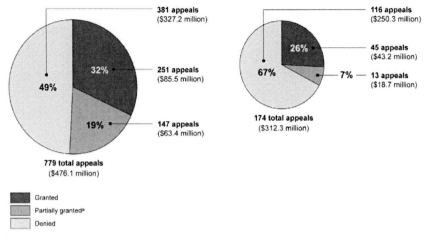

Source: GAO analysis of Federal Emergency Management Agency (FEMA) data. | GAO-18-143.

Note: FEMA data for 2017 includes only first-level appeals that FEMA received between January 1, 2017, and July 12, 2017, and second-level appeals that FEMA received between January 1, 2017, and July 6, 2017. The July 12, 2017, and July 6, 2017, dates are also the end of our period of analysis for the first- and second-level appeals data (respectively) in our review. Dollar figures may not sum to totals due to rounding.

a Amounts in dispute for partially granted appeals do not represent actual award amounts. Rather, for partially granted appeals, FEMA approves only a portion of the relief requested by the applicant.

Figure 5. Outcomes for Appeals FEMA Received between January 2014 and July 2017.

In rendering a final decision on an appeal, FEMA can grant, partially grant, or deny the appeal.[29] Our analysis showed that FEMA granted nearly a third of the 779 first-level appeals filed, awarding applicants over $85 million. As shown in Figure 5, FEMA also partially granted about 19 percent of first-level appeals filed, which involved amounts in dispute totaling over $63 million. Further, Figure 5 shows that over one-third of

---

[29] According to FEMA internal directives, an appeal is considered granted when FEMA has approved the relief requested by the applicant as part of the appeal. An appeal is considered partially granted when FEMA has approved a portion of the relief requested by the applicant. An appeal is considered denied when FEMA has decided not to approve the relief requested by the applicant.

26 *United States Government Accountability Office*

the 174 second-level appeals were either granted or partially granted. Specifically, FEMA granted about 26 percent of second-level appeals filed, awarding over $43 million, while the agency partially granted about 7 percent of second-level appeals filed, involving amounts in dispute totaling almost $19 million.

FEMA Exceeded Statutory Processing Times

Our analysis of the corrected FEMA appeal data showed that, on average, FEMA took more than three times the 90-day statutory time frame to process an appeal, which includes rendering a decision.[30]

Specifically, for first- and second-level appeals that FEMA received between January 2014 and July 2017 and that FEMA decided during the same period, FEMA's average processing time was 297 days. The processing time for decided first-level appeals averaged 293 days, while the processing time for decided second-level appeals averaged 313 days. Further, as shown in Figure 6, only a small percentage of decided first-and second-level appeals (9 and 11 percent, respectively) were processed within the 90-day statutory time frame.

For pending appeals, we found that, at the time of our analysis in July 2017, FEMA had taken on average, more than three times the 90-day statutory time frame for rendering decisions. Specifically, as of July 2017, FEMA had not rendered a decision on 349 appeals, which had an average processing time of 299 days. As of July 2017, the processing time for pending first-level appeals averaged 306 days, while the processing time for pending second-level appeals averaged 267 days. Figure 7 shows the ranges of processing times as of July 2017 for both first-and second level pending appeals.

---

[30] Some appeals may take longer when FEMA issues an RFI. However, according to a PAAB senior official, FEMA uses the 90-day statutory timeframe as a standard for appeal processing timeliness.

Disaster Recovery

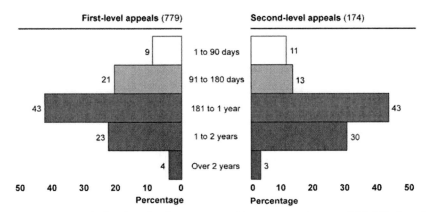

Source: GAO analysis of Federal Emergency Management Agency (FEMA) data. | GAO-18-143.

Note: FEMA data for 2017 includes only first-level appeals that FEMA received between January 1, 2017, and July 12, 2017, and second-level appeals that FEMA received between January 1, 2017, and July 6, 2017. The July 12, 2017, and July 6, 2017, dates are also the end of our period of analysis for the first- and second-level appeals data (respectively) in our review.

Figure 6. Processing Times for Decided Appeals, Based on Appeals FEMA Received between January 2014 and July 2017.

Officials from PAAB and the three regional offices in our review acknowledged that they experienced challenges processing appeals within the 90-day statutory time frame. They told us that issuing RFIs to the applicant can contribute to lengthy processing delays.[31] According to PAAB officials, issuing an RFI may contribute to long processing periods if the information relates to a complex appeal—for example, an appeal involving multiple engineering issues. An appeal decision can also be delayed if FEMA issues an RFI because an applicant submitted incomplete documentation to support an appeal. Under FEMA regulation, these requests do not count against processing times and the 90-day time frame in which FEMA can render a decision on an appeal.[32]

However, our analysis of the corrected FEMA data showed that FEMA exceeded its statutory time frames even when it did not issue an RFI.

---

[31] As discussed earlier in this report, FEMA can issue an RFI to request additional information from an applicant in order to reach a decision.
[32] 44 C.F.R. § 206.206 (c)(3).

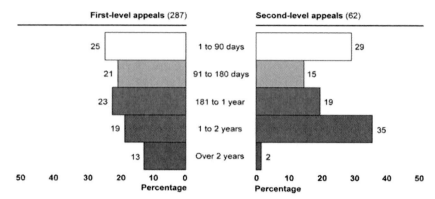

Source: GAO analysis of Federal Emergency Management Agency (FEMA) data. | GAO-18-143.

Note: FEMA data for 2017 includes only first-level appeals that FEMA received between January 1, 2017, and July 12, 2017, and second-level appeals that FEMA received between January 1, 2017, and July 6, 2017. The July 12, 2017, and July 6, 2017, dates are also the end of our period of analysis for the first- and second-level appeals data (respectively) in our review.

Figure 7. Processing Times for Pending Appeals as of July 2017, Based on Appeals FEMA Received between January 2014 and July 2017.

Specifically, between January 2014 and July 2017, FEMA issued an RFI in about 59 percent—or 560—of the 953 first- and second-level appeals for which it rendered a decision. In 48 percent (267) of those decided appeals, FEMA had issued the RFI after the 90-day time frame had elapsed. FEMA did not issue RFIs for about 41 percent (393) of decided first- and second-level appeals. In 78 percent (305) of those appeals, FEMA's processing time still exceeded the 90-day statutory time frame.

State emergency management officials from five of our six selected states told us that they experienced long wait times for first- and second-level appeal decisions and that FEMA rarely processed appeals within the 90- day time frame required by statute. State emergency management officials further told us that such delays adversely affect applicants, such as municipalities and localities, which may wait prolonged periods to resolve project eligibility and costs related to rebuilding efforts.

Delays in FEMA's decision making may also result in additional costs to both the state and the applicant, according to these officials. For example, the state may pursue funding from an applicant if FEMA decides to deobligate funds from the applicant for PA projects that have already been completed.[33] As discussed earlier in this chapter with respect to the PA process, FEMA may do this if it finds that the applicant did not meet certain PA project requirements. In these instances, the applicant may appeal FEMA's decision, but the state may need to begin administrative proceedings against the applicant to recover or offset the deobligated funds. One state emergency manager told us that some applicants withdrew their appeals because of the prolonged delays in receiving a final decision.

According to state emergency management officials, delays in FEMA's appeal decisions can create significant challenges for local government entities, such as counties and school districts. Officials from one state provided an example of a rural school district that sought PA funding to bus displaced children who had been left homeless from damage caused by Hurricane Irene. According to relevant federal and state documents these officials provided us, these children had been moved to shelters outside of their school district and needed transportation to be able to attend school.[34] The school district applied to FEMA for transportation costs associated with hiring an additional bus driver to bus the children to the schools in the district. FEMA denied the school district's request, based on its interpretation of the Stafford Act and the eligibility of costs related

---

[33] As discussed earlier in this report, funds can be deobligated when an agency cancels or downwardly adjusts previously incurred obligations. Specifically, once a DHS OIG audit finds that PA funds were not expended correctly the OIG will make a recommendation to FEMA that the funds be deobligated. If FEMA implements the recommendation, the agency deobligates the funds from the state's PA grant account and the state, in turn, issues a letter and an accompanying invoice to the local government requesting that the previously-approved PA funds be paid back within a certain time frame. In some instances, local governments have agreed to provide full payment to the state, while others have accepted a payment plan to repay the funds over time. Some municipalities have refused to repay the state, resulting in the state's having to off-set any disaster funds that the municipality may request in the future.

[34] The McKinney-Vento Homeless Assistance Act of 1987 is a federal law that provides federal assistance for homeless shelter programs. Pub. L. No. 100-77, 101 Stat. 482, 526- 7 (July 22, 1987). Under the law, a school district is required to transport students whose families obtained temporary housing outside of the school district. 42 U.S.C. § 11432(e)(3)(C)(i)(III(cc), (e)(3)(c)(ii)(II), and (g)(4).

30 *United States Government Accountability Office*

to emergency public transportation. The district subsequently filed a first-level appeal in November 2015. FEMA took over a year to issue a decision and, in December 2016, denied the district's first-level appeal. State management officials told us that incurring these unanticipated transportation costs while waiting for FEMA to decide the appeal has a major effect on the school district and the community as a whole, and can lead to the elimination of school programs or staff. The school district subsequently filed a second-level appeal in February 2017. FEMA denied the appeal in August 2017.

State emergency management officials we interviewed provided an additional example wherein a small town had applied for PA grant funding to rebuild a retaining wall and roadway following damage caused by Hurricane Irene. According to relevant federal and state documents officials provided us, the overflowing banks of a tributary caused a retaining wall, which protected a nearby roadway, to wash away. The roadway, which provided access to residential properties near the tributary, was significantly damaged, due to the overflow. The town requested funding to repair the roadway and to replace and extend the retaining wall another 250 feet beyond the original wall in order to protect the roadway from future flood events. FEMA approved the PA funding to repair the roadway. However, FEMA denied the town's application for PA assistance to extend the wall beyond its original length. In doing so, FEMA concluded that the proposed work was ineligible for assistance because it significantly changed the retaining wall's predisaster configuration and that such a change constituted an improved project, making it ineligible under FEMA regulations and policy. The town then filed a first-level appeal in April 2014. More than 2 years later—in June 2016—FEMA denied the town's first-level appeal, upholding FEMA's original determination. The town subsequently filed a second-level appeal in September 2016. Over a year later, PAAB was still reviewing the appeal.

Disaster Recovery 31

FEMA HAS TAKEN STEPS TO IMPROVE APPEALS PROCESSING, BUT FACES CHALLENGES WITH ITS APPEALS WORKFORCE

FEMA has taken a number of steps to improve its management of the appeals process and respond to issues raised by us and the DHS OIG related to processing delays. As we presented earlier in this chapter, our 2008 review, and DHS's subsequent 2011 OIG review, identified a number of organizational and procedural issues related to processing delays, staff independence, and communications with applicants. Responding to these issues, FEMA created the PAAB within the Recovery Directorate at FEMA Headquarters in late 2013, adding an auditing component to the Branch in 2014.[35] PAAB then established a core of full-time staff at FEMA headquarters that were specifically assigned to process second-level appeals. At the same time, through the Recovery Directorate, each of FEMA's 10 regional offices was assigned full-time staff for processing first-level appeals. Prior to PAAB, second-level appeals were processed by various offices within FEMA headquarters, while the management and processing of first-level appeals varied by FEMA regional office. Some regional offices assigned staff specifically to review appeals, while other offices assigned staff to appeal processing as part of their other responsibilities within the PA Program, such as determining eligibility for PA assistance.[36]

In standing up PAAB, FEMA also established an SOP that describes the organizational structure of PAAB, as well as its responsibilities and the roles of its staff. The SOP also addresses procedures related to PAAB's responsibility for managing the entire PA appeals program. These responsibilities include reporting on appeal processing performance, providing training to appeals staff, and identifying PA appeal process and policy improvements. FEMA later issued a regional SOP that included

---

[35] The Public Assistance Appeals and Audits Branch (PAAB) was originally created as the Public Assistance Appeals Branch in 2013. An audit section was later added to PAAB in 2014.

[36] Department of Homeland Security, Office of Inspector General, *Opportunities to Improve FEMA's Public Assistance Appeals Process*, OIG-11-49 (Washington, D.C.: March 2011).

32   *United States Government Accountability Office*

procedures to help regional offices reduce the number of appeals that exceeded statutory time frames. These procedures reflected an ongoing effort to leverage internal resources when regional offices exceed processing capacity. Specifically, a regional office can submit a request to PAAB for assistance from analyst staff from other regions or from PAAB to assist with processing first-level appeals.[37] PAAB may then temporarily assign an appeals analyst from PAAB or from another regional office to assist the regional office making the request. For example, one regional office official told us his office had requested assistance with 10 first-level appeals and PAAB was able to accommodate the request by assigning 8 of the 10 appeals to another region for processing. According to a senior PAAB official, this procedure allows FEMA to maximize use of its national appeal processing capacity. As of October 2017, PAAB had transferred 77 appeals from overwhelmed regional offices to those with capacity to process additional appeals.

Further, FEMA procedures now require that a conflict check be performed to determine whether the analyst was involved with a PA project determination that is substantively related to the appeal. If a conflict is identified, options include disqualifying the appeals analyst from working on the appeal, or requesting the appeal be transferred to another regional office or PAAB for processing. State emergency management officials from five of the six states in our review told us that they believed that issues related to the independence of appeals staff had been addressed and were no longer an issue.

PAAB also took steps to improve communication with applicants by creating an online second-level appeal tracking spreadsheet—accessible through the Internet—intended to provide applicants with information on the status of second-level appeals.[38] The spreadsheet includes, among other things, the date the appeal was received by FEMA headquarters, the date

---

[37] The Regional PA Branch Chief must email a request to the PAAB Branch Chief. The request should reference the following four items: (1) the level of demand, (2) the existing workforce, (3) the number of appeals the regional office seeks to transfer, and (4) the identification of the time remaining available for an agency response to the applicant.

[38] The FEMA Public Assistance Second-Level Appeals Tracker can be accessed at https://www.fema.gov/media-library/assets/documents/108588.

that an RFI was sent to the applicant, whether the appeal was "under review," whether a final decision had been granted, and the date any final decision was signed.

FEMA also took steps to increase its staffing levels. In January 2015, FEMA's Recovery Directorate completed a workforce analysis and determined that additional appeals analysts were needed to address capacity issues that were resulting in growing inventories of first-level appeals. At the time, FEMA concluded that, in addition to its 23 on-board appeals analysts, an additional 29 appeals analysts were needed to support the existing, as well as anticipated, appeal inventory increases across FEMA's 10 regional offices. The Recovery Directorate requested and was subsequently authorized the additional appeals analyst positions, which, when filled, would provide the PA appeal program with a total of 52 first-level appeals analysts. With the exception of Region I, FEMA planned to provide each of the remaining 9 regional offices with at least 1 additional appeals analyst. Regional offices with the heaviest workloads, such as Region II and Region IV, would be allocated more appeals analysts. FEMA took steps to fill these positions over the next 2 years, and by June 2017, FEMA had filled 47 of the 52 positions.

Despite efforts to improve its management of the appeals process, FEMA faces a backlog of both first- and second-level appeals among the three selected FEMA regional offices as well as PAAB. According to officials in PAAB and the three regional offices in our review, workforce challenges contribute to delays in processing PA appeals, even with the improvements described above.

PAAB and the three regional offices in our review identified the following workforce challenges that contributed to PA appeal processing delays.

Staff Vacancies, Inexperience, and Turnover

Despite FEMA's efforts to increase its appeals analyst staffing level— an effort that began in 2015—two of the three regional offices in our review had a number of vacancies for these positions through June 2017. PAAB and regional officials told us that such vacancies, which occurred over a prolonged period, contributed to appeal processing delays.

FEMA data on appeals analyst staffing show that FEMA took nearly 2 years to fill the additional appeals analyst positions across its 10 regional offices. For example, in 1 of the regional offices in our review, 3 of the 8 appeals analyst positions were vacant through 2016 and were not filled until July 2017. Further, officials in this regional office told us that the current staffing level of 8 appeals analysts was inadequate to keep pace with the region's increasing appeal inventory. Similarly, 6 of PAAB's 11 appeals analyst positions were vacant from August 2015 to October 2016.[39] By July 2017, PAAB had filled all but 2 appeals analyst positions. PAAB officials told us the appeals analyst staffing level consisting of 52 positions was a preliminary estimate and that this staffing level has not been adequate in regions with heavy workloads and appeal inventories. PAAB officials also acknowledged the potential benefits of having an appeals analyst staffing plan, but stated that they are not yet prepared to update the workforce assessment for PAAB and the regional offices, nor do they have plans to do so until full staffing is achieved. These officials also told us that they are still working to achieve the staffing levels developed in 2015 and are taking steps to address staffing challenges through more targeted hiring and use of career ladder positions.

Further, PAAB staffing data showed that almost half of PAAB's staff had less than 1 year of experience. PAAB officials told us that prior vacancies and a large number of inexperienced staff have contributed to processing delays and second-level appeal backlogs. PAAB officials also told us that retaining trained appeals analysts has been challenging due to limited career advancement opportunities within the appeals analyst position. These officials told us that although not required, individuals who

---

[39] These appeals analyst positions included six analysts responsible for processing second-level appeals, and five appeals analysts responsible for processing first-level appeals. According to PAAB policy, these first-level appeals analysts within PAAB coordinate the transfer of first-level appeals from FEMA regional offices to PAAB or to another regional office for processing during such instances as when the originating region has a high appeal volume and cannot process appeal decisions within the 90-day statutory time frame.

Disaster Recovery 35

typically apply for an appeals analyst position possess a law degree, and that once hired, some of them apply for attorney positions within PAAB or in various offices within FEMA or DHS. For example, PAAB staffing data showed that within 18 months of being hired by PAAB, four PAAB appeals analysts applied for and were subsequently hired as attorney-advisors within PAAB or other FEMA departments. Then those appeals analyst positions were vacant until the next round of hiring.

Regional officials told us it has been challenging to find qualified applicants with the specialized skillset of an analyst position. They told us that, ideally, an appeals analyst should be an expert in the PA program and possess a nuanced understanding of the legal issues associated with the program's requirements. Regional officials told us that, because of this specialized skillset, they look to recruit PA appeals analysts from other FEMA regional offices who may have an interest in relocating or are seeking a promotion. However, while recruiting appeals analysts from other regions may assist individual offices, it does not address FEMA's goal of achieving its staffing levels.

Delays in Training Appeals Staff

FEMA requires that PA appeals analysts undergo a certification course that includes 3 days of training on processing appeals. The appeals analyst certification course, delivered through PAAB, covers both procedural steps of processing appeals as well as the policy and legal issues raised by the PA program, and ensures that trainees can prepare a well-written appeal response.[40] After completing the course, an analyst in training must pass a test to demonstrate proficiency in reviewing and analyzing appeals and preparing appeal decisions. To this end, the analyst must analyze a mock appeal—based on facts similar to those presented in a previously decided appeal— and draft an appeal decision.

---

[40] Specific topics covered in the certification course include researching and writing, appeal data management, and developments in PA program policies and laws related to the PA appeal process.

FEMA policy states that only certified staff can serve as appeals analysts and must be recertified every 2 years. However, some appeals analysts in the regional offices in our review had not yet undergone the certification process, but were nonetheless working in an appeals analyst capacity under the supervision of certified analysts. PAAB procedures also state that a trainee analyst cannot assume work on an appeal without being supervised by a certified analyst. For example, in one regional office, four of the office's nine appeals analysts had been working in their positions for between 6 months to a year before they received appeals analyst certification training. According to regional officials, this increased the supervisory workload on the remaining five appeals analysts within the region and the lack of timely training and certification of appeals analysts affect the efficient processing of appeals and can lead to delays in FEMA issuing appeal decisions.

Deployment of Appeals Staff to Disaster Response

According to PAAB officials, while PA appeals analysts are considered "dedicated" positions, these analysts can be deployed at any time to provide assistance on a disaster, such as working with grant applicants to document damages or assisting applicants in developing project proposals to request PA grants. Officials from two of the three FEMA regional offices in our review told us that these deployments contributed to processing delays because, given limited resources, assigning staff to continue work on the appeal is not always possible. In one regional office, five of the nine PA appeals analysts were deployed in late 2016 to do recovery work related to damage from Hurricane Matthew. These deployments lasted approximately 30 to 90 days and left the regional office understaffed. Further, one regional office official told us that maintaining continuity in processing an appeal can be difficult for those analysts who are deployed because they must pick up where they left off on their assigned appeals upon their return.

A senior PAAB official told us that regional appeals analyst staff have been deployed to assist with response and recovery efforts as a result of the catastrophic damage from Hurricanes Harvey, Irma, and Maria. As a result, these analysts have not been available to process first-level appeals. This official further told us that PAAB staff, including analyst staff—while not deployed—have been assigned to support disaster operations. For example, one staff member was assigned to support site inspector training, while two others were assigned to stand National Response Coordination Center watch.[41] Further, one staff member was assigned to support training and contract review functions and the remaining staff members were assigned as call takers for the PA Grants Manager and Grants Portal hotline.[42]

To help overcome staffing shortages, according to FEMA documents, all three regional offices in our review staffed assistance from PAAB at various times during the past 2 years. However, officials from two of the three regional offices in our review told us that, based on their experiences, requesting staff from PAAB or other offices had a number of limitations. Specifically, because the originating regional office is ultimately responsible for the appeal, its staff must continue to oversee the appeal, including such responsibilities as tracking the appeal, corresponding with the applicant and the state as needed, and reviewing and approving the appeal decision. One regional office official told us that this arrangement was not helpful and only added an additional layer of complexity that delayed processing. Another regional official told us that the quality of the borrowed staff's work was not consistent. This official further stated that, because offices are not able to select the analysts that would be assigned to

---

[41] The National Response Coordination Center (Center) is a multiagency center that coordinates the overall federal support for major incidents and emergencies. The Center coordinates with the affected region(s) and provides resources and policy guidance in support of the incident. The Center staff consists of FEMA personnel, appropriate Emergency Support Functions from various federal agencies, and other appropriate personnel and agencies.

[42] The PA Grants Manager and Grants Portal is a web-based project tracking and case management tool designed to formalize standard processes and provide applicants with real-time data and information on project status. The system is also designed to provide the ability to capture applicant documents, maintain applicant and disaster profiles, and improve automated reporting. The PA Grants Manager is accessible to FEMA employees while the PA Grants Portal is accessible to applicants.

38 *United States Government Accountability Office*

work on their appeals, he was reluctant to use staff from other regional offices.

According to leading human capital practices, the key to an agency's success in managing its programs is sustaining a workforce with the necessary knowledge, skills, and abilities to execute a range of management functions that support the agency's mission and goals.[43] Achieving such a workforce depends on having effective human capital management through developing human capital strategies. Such strategic workforce planning includes the agency assessing current and future critical skill needs by, for example, analyzing the gaps between current skills and future needs, and developing strategies for filling the gaps identified in workforce skills or competencies. *Standards for Internal Control in the Federal Government* also states that agencies should continually assess their needs so that they are able to obtain a workforce that has the required knowledge, skills, and abilities to achieve their organization's goals.[44] Further, as we have previously reported in our work on strategic workforce planning, such staffing assessments should be based on valid and reliable data.[45]

However, FEMA has not developed a workforce staffing plan to identify hiring, training, and retention needs of appeals staff across PAAB and the regional offices. PAAB officials told us that they are still working to achieve the staffing levels developed in 2015 and are taking steps to address staffing challenges related to retention through more targeted hiring and use of career ladder positions. In the absence of a workforce plan for the PA appeals staff, FEMA will likely continue to experience workforce challenges including vacancies in key appeals analyst positions, appeals staff turnover, training delays, and understaffing due to disaster deployment. These challenges will likely continue to contribute to delays

---

[43] GAO, *Federal Emergency Management Agency: Workforce Planning and Training Could Be Enhanced by Incorporating Strategic Management Principles,* GAO-12-487 (Washington, D.C.: Apr. 26, 2012) and *Human Capital: Key Principles for Effective Strategic Workforce Planning,* GAO-04-39 (Washington, D.C.: Dec. 11, 2003).

[44] GAO-14-704G.

[45] GAO-12-487.

Disaster Recovery

in FEMA's processing and issuing first- and second-level PA appeals decisions.

FEMA ESTABLISHED GOALS AND MEASURES TO ASSESS SECOND-LEVEL APPEAL PROCESSING, BUT DID NOT DO SO FOR FIRST-LEVEL APPEALS

FEMA officials have acknowledged the importance of establishing goals and measures to assess the performance of the PA appeals program. In particular, for fiscal year 2016, FEMA's Recovery Directorate established two performance goals for PAAB's processing of second-level appeals. The first goal was aimed at reducing the inventory of second-level appeals by 20 percent. The second goal was aimed at processing at least 30 percent of second-level appeals received in 2016 within 90 days of receiving the appeal, in order to comply with FEMA statutory time frames.[46] FEMA internal documents showed that these two performance goals were intended to reduce the second-level appeal inventory, and, at the same time, promote a standard of timely second-level appeal processing for PAAB.

According to PAAB officials, various factors beyond PAAB's control prevented PAAB from meeting these performance goals. These factors included an unanticipated surge in the number of second-level appeals in 2016, as well as increased vacancies due to staff turnover in PAAB analyst positions in 2016. Recognizing these factors, PAAB developed a revised goal that focused on the number of appeals an analyst could process per month.[47] According to PAAB officials, focusing the revised goal on analyst production controlled for external factors that tended to affect

---

[46] In addition, a third performance goal included issuing the first Regional Appeals Standard Operating Procedures as well as updating the PA Program Appeals Procedures Directive and Manual to reflect current appeal processes and regulatory changes.

[47] PAAB documentation showed that, for purposes of performance evaluation, to meet expectations the issuance rate for an appeals analyst is 0.7 appeals per month and 1 per month to exceed expectations.

overall processing, such as surges in appeal submissions and staff turnover. PAAB officials told us that their proposed production goal was not accepted by the Recovery Directorate for 2016, but that PAAB adopted the revised goal for individual performance plans for PAAB appeals analyst staff.

In contrast, although first-level appeals represent the majority of FEMA's appeal inventory, FEMA has not developed goals and measures to assess the performance of first-level appeals processing across regional offices. PAAB collects various data from all 10 regional offices on first-level appeals, such as the number of first-level appeals being processed, as well as processing timeliness (i.e., appeals that exceeded time limits) and key words that can help identify various appeal subject-matter categories. PAAB then aggregates this data, which it publishes on a quarterly and weekly basis in internal reports that it shares with FEMA management. However, FEMA has not established goals to assess performance against the information that PAAB collects. According to FEMA officials, while the Recovery Directorate established goals and measures for second-level appeals, it is not responsible for developing goals and measures to assess performance within the regional offices. These officials told us further that some Regional Administrators have established goals and measures for first-level appeals within their regional offices, while others have not.

For management to effectively monitor a program, *Standards for Internal Control in the Federal Government* state that it should create goals and measures to determine if a program is being implemented as intended. In addition, the quality of the program's performance should be assessed over time and monitoring efforts should be evaluated to assure they help meet goals.[48] Further, Congress enacted the GPRA Modernization Act of 2010 (GPRAMA) to focus and sustain attention on agency performance and improvement by requiring that federal agencies establish outcome-oriented goals and measures to assess progress towards those goals.[49] Specifically, agencies, like DHS, are required to monitor

---

[48] GAO-14-704G.

[49] Pub. L. No. 111-352, 124 Stat. 3866 (Jan. 4, 2011). GPRAMA significantly enhanced the Government Performance and Results Act of 1993 (GPRA) by providing important tools

progress towards the achievement of goals, report on that progress, and address issues identified. Without consistent performance measures across FEMA regional offices to help assess progress and identify deficiencies in appeals processing, DHS and its subcomponent agencies like FEMA may have difficulty providing accurate reporting on the effectiveness of current efforts to process first-level appeals and on the factors that contribute to ongoing appeal processing delays.

CONCLUSION

Although FEMA has made efforts to improve its management of the PA appeals process, these efforts have been hampered by a number of issues including weaknesses in FEMA's appeals tracking data and its ability to ensure the quality of this data. FEMA corrected its appeals data for purposes of this chapter once we pointed out data discrepancies, but FEMA does not have a process to ensure data quality issues are permanently addressed. As a result, these weaknesses will persist. By implementing procedures to consistently track appeals data and ensure the quality of these data, FEMA will be in a better position to accurately report on appeal processing performance and make informed decisions about the appeals process.

FEMA also faces a variety of workforce challenges that have contributed to appeals processing delays. These challenges include staffing vacancies, lack of experienced staff, high rates of staff turnover, delays in training appeals staff, and the deployment of appeals analysts for disaster response, all of which have contributed to processing delays. Addressing these challenges by identifying the hiring, training, and retention needs of its appeals offices through strategic workforce planning could help FEMA better position itself to reduce its appeals backlog and better respond to PA appeals.

---

that can help agencies resolve their major management challenges. GPRA, Pub. L. No. 103-62, 107 Stat. 285 (Aug. 3, 1993).

Further, although FEMA has established goals and measures for its second-level appeals processing, it has not done so for first-level appeals. By establishing goals and measures to assess the performance of its first-level appeals process, DHS and FEMA will be able to better evaluate the efficiency and effectiveness of its efforts to reduce the PA appeal backlog and improve appeal processing times.

Recommendations for Executive Action

We are making the following four recommendations to FEMA:

The Assistant Administrator for Recovery should design and implement the necessary processes and procedures to ensure a uniform and consistent approach for tracking first-level appeals data to better integrate regional trackers with PAAB's own first-level appeals tracker. (Recommendation 1)

The Assistant Administrator for Recovery should design and implement the necessary controls to ensure the quality of the first-level appeals data collected at and reported from the regional offices to PAAB. (Recommendation 2)

The Assistant Administrator for Recovery should develop a detailed workforce plan that documents steps for hiring, training, and retaining key appeals staff. The plan should also address staff transitions resulting from deployments to disasters. (Recommendation 3)

The Assistant Administrator for Recovery should work with Regional Administrators in all 10 regional offices, to establish and use goals and measures for processing first-level PA appeals to monitor performance and report on progress. (Recommendation 4)

AGENCY COMMENTS AND OUR EVALUATION

We provided a draft of this chapter to the Secretary of the Department of Homeland Security and the Administrator of the Federal Emergency

Disaster Recovery

Management Agency for review and comment. DHS provided written comments, which are reproduced in appendix II. In its comments, DHS concurred with our recommendations and described actions planned to address them. FEMA also provided technical comments, which we incorporated as appropriate. Additionally, we provided excerpts of the draft report to state emergency management officials in the selected six states we included in our review. We incorporated their technical comments as appropriate.

Regarding our first recommendation, that FEMA design and implement the necessary processes and procedures to ensure a uniform and consistent approach for tracking first level-appeal data, DHS stated that FEMA's PAAB will develop guides and checklists for the regions to ensure data uniformity and consistency and that PAAB will update its data review process, and develop additional content highlighting the importance of data integrity and accuracy. DHS estimated that this effort would be completed by July 31, 2018.

Regarding our second recommendation, that FEMA design and implement the necessary controls to ensure first-level appeal data quality, DHS stated that PAAB will include content within the certified appeal analyst training highlighting the importance of data integrity and that first-level appeal data will be reviewed by PAAB on a quarterly basis. DHS estimated that this effort would be completed by February 28, 2019.

Regarding our third recommendation, that FEMA develop a detailed workforce plan for hiring, training and retaining key appeals staff, DHS stated that by December 31, 2018, PAAB will produce a workload flow assessment on second-level appeals staffing and determine whether appeal timeliness issues still exist. If PAAB determines that significant response timeliness issues on second-level appeals still exist after most PAAB appeal analyst staff have at least one year of experience, a detailed PAAB workforce plan will be completed and finalized by December 31, 2019. PAAB will also complete an assessment of first-level appeal inventory and timeliness issues. If PAAB determines that significant regional response inventory and timeliness issues on first-level appeals still exist, FEMA will create a working group to prepare a detailed regional workforce plan. DHS estimated that this effort would be completed by December 31, 2019.

Regarding our fourth recommendation that FEMA work with Regional Administrators to establish and use performance goals and measures for processing first-level appeals, DHS stated that PAAB has begun developing a methodology for establishing, measuring, and reporting on first-level appeals processing goals and performance progress, and that PAAB would work with the regions to complete and finalize this methodology. DHS estimated that this effort would be completed by August 31, 2018.

As agreed with your offices, unless you publicly announce the contents of this chapter earlier, we plan no further distribution until 30 days from the report date. At that time, we are sending copies of this chapter to the Secretary of Homeland Security and interested congressional committees.

Allison B. Bawden
Director, Strategic Issues

List of Requesters

The Honorable Ron Johnson
Chairman

The Honorable Claire McCaskill
Ranking Member
Committee on Homeland Security and Governmental Affairs
United States Senate

The Honorable Thomas R. Carper
Ranking Member
Permanent Subcommittee on Investigations
Committee on Homeland Security and Governmental Affairs
United States Senate

The Honorable Bill Shuster
Chairman

The Honorable Peter DeFazio
Ranking Member
Committee on Transportation and Infrastructure
House of Representatives

APPENDIX I: OBJECTIVES, SCOPE, AND METHODOLOGY

This chapter reviews aspects of the Federal Emergency Management Agency's (FEMA) management of the Public Assistance (PA) appeals process. The objectives of this review were to determine: (1) the extent to which FEMA ensures quality in its data on appeals and what FEMA data show about its appeals inventory and timeliness for appeals decisions; (2) what steps FEMA has taken to improve its management of the appeals process and what challenges, if any, remain; and (3) the extent to which FEMA has developed goals and measures to assess the appeal program's performance.

To address the first objective, we obtained and analyzed data from FEMA on all first- and second-level appeals that the agency received between January 2014 and July 2017. For first-level appeals, FEMA provided us data on appeals received between January 1, 2014, and July 12, 2017, while FEMA provided us data on second-level appeals received between January 1, 2014, and July 6, 2017. We focused on this time frame because it contained the most complete and available data on each type of appeal at the time of our review. We identified various discrepancies in the first-level appeals data, which we discussed with knowledgeable FEMA staff. Examples of these discrepancies, which we present in this chapter, included missing data, erroneous data entries, and inconsistent recording of data. In response to our discussions, FEMA provided us with corrected data to address the identified discrepancies.

46   *United States Government Accountability Office*

After obtaining the corrected data, we concluded the appeals data from FEMA were sufficiently reliable to provide information on PA appeals that we present in this chapter. We also obtained and analyzed FEMA policies and procedures related to tracking appeals data, such as FEMA's policies and procedures related to regional offices, and evaluated them using *Standards for Internal Control in the Federal Government.*[50]

We analyzed the corrected data to determine FEMA's appeal inventory— that is, the number of first-and second-level appeals that were pending and decided, including any amounts in dispute or amounts awarded, and appeal outcomes for appeals that FEMA decided. From the total number of appeals received, we excluded four second-level appeals that had been remanded or rescinded.[51] We determined the processing times for first- and second-level decided appeals by calculating, for each appeal, the number of calendar days between the date that FEMA received the appeal and the date that FEMA rendered a decision on the appeal.

We then calculated the average number of calendar days to determine average processing times for first- and second-level decided appeals. We determined the processing time for pending first-level appeals by calculating, for each appeal, the number of calendar days between the date FEMA received the appeal and July 12, 2017. Similarly, we determined the processing time for pending second-level appeals by calculating, for each appeal, the number of calendar days between the date FEMA received the appeal and July 6, 2017. We then calculated the average number of calendar days to determine average processing times for pending first-and second-level appeals. We compared processing times for first- and second-level appeals against FEMA's 90-day statutory time frame to determine the number of calendar days by which FEMA exceeded the time frame.

---

[50] GAO, *Standards for Internal Control in the Federal Government*, GAO-14-704G (Washington, D.C.: September 2014).

[51] According to FEMA, a rescinded appeal occurs when a FEMA regional office has made a procedural error during the first-level appeal process (e.g., failing to properly close the administrative record) and the Regional Administrator opts to correct the matter by reopening the first-level appeal process.

Disaster Recovery

47

We also determined the number of first- and second-level appeals in which FEMA issued an RFI and those in which FEMA did not issue an RFI. For the first- and second-level appeals in which FEMA issued an RFI, we compared the date the appeal was received to the date that FEMA issued the RFI. We used the first RFI in cases where FEMA issued multiple RFIs. We then determined whether FEMA had issued the RFI within 90 calendar days. For the first- and second-level appeals in which FEMA did not issue an RFI, we compared the date the appeal was received to the date that FEMA issued a decision. We then determined whether FEMA had issued a decision after 90 calendar days. We also obtained and analyzed FEMA policies and procedures and program directives governing appeal data collection and evaluated them against *Standards for Internal Control in the Federal Government.*[52]

To address the first and second objectives, we also administered semistructured interviews to officials from 3 of FEMA's 10 regional offices (Regions II, IV, and VI) with the highest number of first- and second-level pending appeals. We asked these officials about their efforts to process and track appeals, what improvements had been made regarding how PA appeals are processed, as well as what challenges they believed remained in processing PA appeals since 2013.To select these offices, we obtained data from FEMA on first- and second-level appeals that were pending a decision, as of October 31, 2016. Collectively, these appeals represented 69 percent of all pending first- and second-level appeals FEMA had received as of October 31, 2016. We focused on this time frame because it contained the most recent data for selecting FEMA regional offices at the time of our review. To obtain additional perspective on what, if any, challenges remain in FEMA's management of the appeals process, we also interviewed state emergency management officials in six states (two states in each of the corresponding 3 FEMA regional offices). (See Table 2).

---

[52] GAO-14-704G.

48　　　*United States Government Accountability Office*

The information obtained from the FEMA regional offices and the state emergency management offices cannot be generalized nationwide. However, the information obtained from these officials provides insight into the issues FEMA encountered during the appeal process.

Table 2. Number of pending first- and second-pending appeals by FEMA region and state, as of October 31, 2016

FEMA Region	State	Pending First-Level appeals	Pending Second-Level appeals	Total
II	New Jersey	23	2	25
	New York	15	8	23
IV	Florida	90	11	101
	Mississippi	28	2	30
VI	Louisiana	13	23	36
	Texas	63	7	70
Total		232	53	285

Source: GAO analysis of Federal Emergency Management Agency (FEMA) data. I GAO-18-143.

To additionally address the second objective, we reviewed our past report and Department of Homeland Security Inspector General reports on the PA appeals program. We also reviewed FEMA documentation, such as policy directives, internal staffing requests, appeals analyst position descriptions, and other internal memoranda. We used these sources to identify what steps FEMA had taken to improve its management of the appeals process since 2013. We also used this information to supplement our understanding of the challenges the Public Assistance Appeals Board (PAAB) and regional officials raised during our interviews discussed above.

To address the third objective, we analyzed a series of FEMA internal performance reports issued between November 29, 2013, and February 15, 2017. Developed by PAAB and provided to FEMA management on a quarterly basis, these reports included aggregate information on PA appeals inventory, such as the number of first- and second-level pending appeals, the number of appeals processed within statutory timeframes, the number of pending appeals that are beyond the statutory timeframe, and common appeal issues based on keywords entered by analysts responsible for processing appeals. We also analyzed internal documents, such as briefs and newsletters, which provided detail on specific appeal decisions as well as the status of the appeals inventory. Further, we analyzed FEMA's Strategic Plans for fiscal years 2008 to 2013 and fiscal years 2014 to 2018 to identify objectives, measures, and overall agency-wide goals. We assessed the information in these documents against leading practices in measuring agency performance[53] and against federal standards for internal control.[54]

For all three objectives, we reviewed relevant legislation and FEMA standard operating procedures that govern both FEMA headquarters and regional offices. We also interviewed officials in PAAB and FEMA's Recovery Directorate.

APPENDIX II: COMMENTS FROM THE DEPARTMENT OF HOMELAND SECURITY

U.S. Department of Homeland Security
Washington, DC 20528

December 6, 2017

Allison B. Bawden
Director, Strategic Issues
U.S. Government Accountability Office
441 G Street, NW
Washington, DC 20548

Re: Management's Response to Draft Report GAO-18-143, "DISASTER RECOVERY: Additional Actions Would Improve Data Quality and Timeliness of FEMA's Public Assistance Appeals Processing"

---

[53] GAO, *Managing for Results: Agencies' Trends in the Use of Performance Information to Make Decisions*, GAO-14-747 (Washington, D.C.: Sept. 26, 2014) and GAO, *Executive Guide: Effectively Implementing the Government Performance and Results Act*, GAO/GGD-96-118 (Washington, D.C.: June 1996). While GPRA is applicable to the department or agency level, performance goals and measures are important management tools applicable to all levels of an agency, including the program, project, or activity level, consistent with leading management practices and internal controls related to performance monitoring.

[54] GAO-14-704G.

50 *United States Government Accountability Office*

Dear Ms. Bawden:

Thank you for the opportunity to review and comment on this draft report. The U.S. Department of Homeland Security (DHS) appreciates the U.S. Government Accountability Office's (GAO) work in planning and conducting its review and issuing this report.

The Department is pleased to note the GAO's recognition of improvements the Federal Emergency Management Agency (FEMA) has made in managing the Public Assistance (PA) appeals process, including increasing its appeals staffing levels and developing standard operating procedures. For example, each of FEMA's 10 regional offices now has full-time staff dedicated to processing first-level appeals. FEMA is committed to (1) ensuring it issues consistent appeal decisions within legally-mandated response times, and (2) identifying areas where it can improve the PA program policies and procedures. The later includes enhancing communications with the grantees and applicants and further enabling field staff to render high-quality, accurate eligibility determinations in order to decrease the number of appeals filed by applicants.

The draft report contained four recommendations with which the Department concurs. Attached find our detailed response to each recommendation.

Again, thank you for the opportunity to review and comment on this draft report. Technical comments were provided under a separate cover. Please feel free to contact me if you have any questions. We look forward to working with you again in the future.

Sincerely,

JIM H. CRUMPACKER, CIA, CFE
Director
Departmental GAO-OIG Liaison Office

Attachment

Attachment: Management Response to Recommendations Contained in GAO-18-143

GAO recommended that FEMA's Assistant Administrator for Recovery:

Recommendation 1: Design and implement the necessary processes and procedures to ensure a uniform and consistent approach for tracking first-level appeals data to better integrate regional trackers with PAAB [Public Assistance Appeals Branch's] own first-level appeals tracker.

Response: Concur. The FEMA Recovery Directorate's PAAB will design and implement necessary processes and procedures to ensure a uniform and consistent approach for tracking first-level appeals data. This effort will include development of guides and checklists for the regions to ensure data uniformity and consistency as each region inputs data in PAAB's first-level appeals SharePoint tracker. PAAB will also update its first-level appeals SharePoint tracker data review process to ensure all mandatory data fields are completed by regions, or develop a mechanism to give a warning for any blank data fields as regions input data. Additionally, PAAB will develop and include additional content highlighting the importance of data integrity and accuracy in the certified appeal analyst training material. Estimated Completion Date (ECD): July 31, 2018

Recommendation 2: Design and implement the necessary controls to ensure the quality of the first-level appeals data collected at and reported from the regional offices to PAAB.

Response: Concur. PAAB will include content within the certified appeal analyst training that highlights the importance of data integrity, by July 31, 2018. PAAB will also notify, by September 30, 2018, FEMA regions that the first-level appeal SharePoint tracker will be reviewed on a quarterly basis (starting with a review of data collected during the fourth quarter of calendar year 2018) and that the results of these reviews will be shared with all regions. ECD: February 28, 2019

Disaster Recovery

Recommendation 3: Develop a detailed workforce plan that documents steps for hiring, training, and retaining key appeals staff. The plan should also address staff transitions resulting from deployments to disasters.

Response: Concur. By December 31, 2018, PAAB (1) will produce a work flow assessment on second appeal staffing to optimize appeals staffing levels and determine whether second appeal response timeliness issues still exist, and (2) will complete an assessment on first appeal regional inventory and timeliness issues. If significant response timeliness issues on second appeals are determined to still exist after a majority of PAAB appeal analyst staff have attained at least one year of experience, a detailed PAAB workforce plan will be completed and finalized by December 31, 2019. If significant regional response inventory and timeliness issues on first appeals are determined to still exist, FEMA will then create a working group, including representatives from all 10 regions, to prepare and finalize a detailed regional workforce plan. ECD: December 31, 2019.

Recommendation 4: Work with Regional Administrators in all 10 regional offices to establish an use goals and measures for processing first-level PA appeals to monitor performance and report on progress.

Response: Concur. PAAB has begun developing the methodology to be used for establishing, measuring, and reporting on first-level appeals processing goals and performance progress. PAAB will work with the regions to complete and finalize the methodology; afterwards, it will be up to regions to monitor their performance. ECD: August 31, 2018

In: Issues in Disaster Recovery and Assistance ISBN: 978-1-53616-308-7
Editor: Donatien Moïse © 2019 Nova Science Publishers, Inc.

Chapter 2

FEMA INDIVIDUAL ASSISTANCE PROGRAMS: IN BRIEF*

Shawn Reese

SUMMARY

When the President declares a major disaster pursuant to the Robert T. Stafford Disaster Relief and Emergency Assistance Act (P.L. 93-288), the Federal Emergency Management Agency (FEMA) advises the President about types of federal assistance administered by FEMA available to disaster victims, states, localities, and tribes. The primary types of assistance provided under a major disaster declaration include funding through the Public Assistance program, Mitigation Assistance programs, and the Individual Assistance program.

The Public Assistance program provides federal financial assistance to repair and rebuild damaged facilities and infrastructure. Mitigation Assistance programs provide funding for jurisdictions, states, and tribes to ensure damaged facilities and infrastructure are rebuilt and reinforced to better withstand future disaster damage. Finally, the Individual

---

* This is an edited, reformatted and augmented version of Congressional Research Service Publication No. R45085, Updated January 31, 2018.

Assistance program provides funding for basic needs for individuals and households following a disaster.

Eligible activities under the Individual Assistance program include funding for such things as mass care, crisis counseling, and temporary housing. FEMA advises the President on the type of individual assistance to be granted following each disaster, and works with state and local authorities in determining what assistance programs would best suit the needs within the disaster area. FEMA makes this determination based on a list of criteria designed to align federal disaster assistance with unmet needs in disaster-impacted areas.

This chapter provides a short summary of the types of individual assistance programs administered by FEMA following a disaster. This chapter also provides a summary of the criteria FEMA uses in determining which individual assistance programs may be made available to impacted areas following a major disaster declaration, and discusses a proposed rule to change these criteria.

INTRODUCTION

When the President declares a major disaster under the Robert T. Stafford Disaster Relief and Emergency Assistance Act,[1] the disaster declaration designates the types of Federal Emergency Management Agency (FEMA) assistance to be provided.[2] There are three primary types of assistance: Public Assistance (PA),[3] which addresses repairs to a community and state or tribal government infrastructure; Mitigation Assistance (MA), which provides funding for projects a state or tribe proposes to reduce the threat of future damage; and Individual Assistance (IA), which provides help to individuals and households affected by a major disaster.

While all major disaster declarations include MA, FEMA uses information on a range of "factors" from the Preliminary Damage Assessment (PDA) to determine whether PA and/or IA should be

---

[1] 42 U.S.C. §5170, as amended.

[2] For more information on Stafford Act declarations, see CRS Report R43784, *FEMA's Disaster Declaration Process: A Primer*, by Bruce R. Lindsay.

[3] For more information on PA, see CRS Report R43990, *FEMA's Public Assistance Grant Program: Background and Considerations for Congress*, by Jared T. Brown and Daniel J. Richardson.

FEMA Individual Assistance Programs: In Brief 55

recommended.[4] Following Hurricane Sandy, Congress required FEMA to revise and update the factors it considers when making a recommendation to the President regarding whether a major disaster declaration should include IA or not.[5] This chapter provides information on FEMA's IA programs and the factors FEMA uses to determine if IA should be part of a disaster declaration.

INDIVIDUAL ASSISTANCE PROGRAMS

FEMA considers numerous factors (detailed later in this chapter) when it recommends IA to the President following a major disaster. IA can include several programs, depending on whether the governor of the affected state or the tribal leader requests that specific type of FEMA assistance. FEMA's IA includes (1) Mass Care and Emergency Assistance, (2) Crisis Counseling Assistance and Training Program, (3) Disaster Unemployment Assistance, (4) Disaster Legal Services, (5) Disaster Case Management, and (6) the Individuals and Households Program.[6]

1. *Mass Care* includes directly supporting congregate sheltering, feeding, and hydration; distributing emergency supplies; and reuniting children with their parents/legal guardians, as well as adults with their families. *Emergency Assistance* encompasses a variety of services and functions, including coordination of volunteer organizations and unsolicited donations, managing unaffiliated volunteers and community relief services, supporting transitional sheltering, and supporting mass evacuations.

---

[4] In general, when a request for a major disaster declaration is submitted, representatives from FEMA meet with the state or tribal government and compile a PDA. FEMA then makes a recommendation to the President concerning whether a declaration should be issued. The President has the authority to make the declaration or deny the request. For more information on the PDA process, see CRS Report R44977, *Preliminary Damage Assessments for Major Disasters: Overview, Analysis, and Policy Observations*, coordinated by Bruce R. Lindsay.

[5] P.L. 113-2, Division B.

[6] P.L. 93-288, Title IV, §401-426.

2. The *Crisis Counseling Assistance and Training Program* assists individuals and communities recovering from the effects of a disaster through community-based outreach and psycho-educational services.[7] The program supports short-term counseling of disaster survivors. The program also provides information on coping strategies and emotional support by linking survivors with other individuals and agencies that help them in the recovery process.

3. *Disaster Unemployment Assistance* provides information and resources to individuals who were employed or self-employed, or were scheduled to begin employment during a disaster. It may also be provided to those who can no longer work or perform their job duties due to damage to their place of employment, do not qualify for regular unemployment benefits from a state, or cannot perform work or self-employment due to an injury as a direct result of a disaster.[8]

4. *Disaster Legal Services* provides legal assistance to low-income individuals who are unable to secure adequate legal services that meet their disaster-related needs.

5. *Disaster Case Management* provides a partnership between a case manager and the disaster survivor to assist them in carrying out a disaster recovery plan. The recovery plan includes resources, services, decisionmaking priorities, progress reports, and the goals needed to close their case.

6. *Individuals and Households Program* is comprised of two categories of assistance: Housing Assistance and Other Needs Assistance (ONA).

   a. Housing Assistance may include financial assistance to
      i   reimburse for hotels, motels, or other short-term lodging;
      ii rent alternate housing accommodations while the applicant is displaced from their primary residence;
      iii repair a primary residence;

---

[7] Psycho-educational services consist of therapeutic treatment for disaster victims that provides information and support to help them better understand and cope with their situation.

[8] For more information on disaster unemployment assistance, see CRS Report RS22022, *Disaster Unemployment Assistance (DUA)*, by Julie M. Whittaker.

FEMA Individual Assistance Programs: In Brief 57

iv assist in replacing owner-occupied residences when the residence is destroyed; and

v enter into lease agreements with owners of multifamily rental properties located in the disaster area.[9]

b. Housing Assistance may also include home repair and construction services provided in insular areas outside the continental United States and other locations where no alternative housing resources are available and where types of FEMA housing assistance that are normally provided (such as rental assistance) are unavailable, infeasible, or not cost-effective.

c. FEMA may provide manufactured housing units as a form of temporary housing through its Transitional Sheltering Assistance program.

d. Other Needs Assistance provides financial assistance for other disaster-related expenses and needs. These include

i child care;

ii medical and dental expenses;

iii funeral and burial costs; and

iv transportation.

IA accounted for approximately 14.4% of all projected federal spending ($9.1 billion of $63.2 billion) from the Disaster Relief Fund for Stafford Act declarations occurring from FY2007 through FY2016, according to FEMA data as of August 10, 2017. Some types of FEMA IA are subject to cost sharing, such as ONA, which is subject to a cost share between FEMA and the state, territorial, or tribal government. FEMA covers 75% of eligible ONA costs, and the state, territorial, or tribal government is responsible for the remaining 25%.[10] Transitional Sheltering under Mass Care and Emergency Assistance is also subject to a 75/25 percent cost share.

---

[9] Housing assistance is capped at $33,300 per household.

[10] 42 U.S.C. §5174(g) and 44 C.F.R. §206.110(i).

IA Factors for a Major Disaster Declaration

The factors FEMA uses to determine potential IA for disaster survivors have not been changed since originally published in 1999. The factors include

- *concentration of damages*—FEMA evaluates the concentrations of damages to individuals, and high concentrations of damages generally indicate a greater need for federal assistance than widespread or scattered damages throughout an area, region, or state;
- *trauma*—FEMA considers the degree of trauma to a state and communities, and some of the conditions that might cause such trauma include
  - o large numbers of injuries and death,
  - o large-scale disruption of normal community functions and services, and
  - o emergency needs such as extended or widespread loss of power or water;
- *special populations*—FEMA considers whether special populations, such as low-income, the elderly, or the unemployed are affected, and whether they may have a greater need for assistance;
- *voluntary agency assistance*—FEMA considers the extent to which voluntary agencies and state or local programs can meet the needs of the disaster victims;
- *insurance*—FEMA considers the amount of insurance coverage in the disaster area because, by law,[11] federal disaster assistance cannot duplicate insurance coverage; and
- *average amount of individual assistance by state*—FEMA's regulations state that "[t]here is no set threshold for recommending Individual Assistance, but the following averages [see *Table 1*]

---

[11] 42 U.S.C. §5155.

FEMA Individual Assistance Programs: In Brief

may prove useful to States and voluntary agencies as they develop plans and programs to meet the needs of disaster victims."[12]

**Table 1. Average Amount of Assistance Per Disaster
July 1994-July 1999**

	Small states <2M pop.	Medium states 2-10M pop.	Large states >10M pop.
Average population (1990 census data)	1.0M	4.7M	15.5M
Disaster Housing Applications Approved	1,507	2,747	4,679
Homes Estimated Major Damage/Destroyed	173	582	801
Housing Assistance	$2.8M	$4.6M	$9.5M
Individual and Family Grant (IFG) Applications Approved	495	1,377	2,071
Individual and Family Grant Assistance	$1.1M	$2.9M	$4.6M
Disaster Housing/IFG Combined Assistance	$3.9M	$7.5M	$14.1M

Source: Data from FEMA's National Processing Service Centers, as presented in 44 C.F.R. §206.48(b).
Note: Data used are only from July 1994 to July 1999. M = millions.

When reviewing disaster declarations between January 2008 and July 2013, FEMA found that requests with more than $7.5 million in assistance usually resulted in a declaration for IA. However, FEMA has not suggested using that amount as a threshold for assistance. Instead, it is providing several factors that relate directly to the Individual and Household Program (IHP), which is the primary IA program. In addition, other factors may provide greater detail for decisionmakers for IHP and for other prominent IA programs such as Crisis Counseling and Disaster Unemployment Assistance (DUA).

In response to congressional direction[13] for updated IA factors, FEMA issued a Notice of Proposed Rulemaking (NPRM) on November 12, 2015. FEMA published the proposed rule on November 12, 2016, and was accepting comments until January 11, 2016. FEMA has yet to finalize the proposed rule. The new proposed factors are

---

[12] 44 C.F.R. §206.48(b).
[13] P.L. 113-2, Division B.

60   Shawn Reese

- state fiscal capacity and resource availability;
- uninsured home and personal property losses;
- disaster-impacted population profile;
- impact to community infrastructure;
- casualties; and
- disaster-related unemployment.

FEMA proposed using three factors to evaluate a state and local jurisdiction's fiscal capacity: the total tax revenue of the state (a measurement recommended by the U.S. Government Accountability Office [GAO]),[14] the Gross Domestic Product (GDP) by state, and the per-capita personal income by local area. Using those factors, along with the projected IA costs, FEMA would then develop Individual Assistance Cost to Capacity (ICC) ratios.

As mentioned earlier, the estimates of potential FEMA assistance costs are generally derived from Preliminary Damage Assessments (PDAs) that are conducted after an event to help both the state and federal governments evaluate the situation. Using figures from this process for 153 major disaster declaration requests between January of 2008 and July of 2013, FEMA found that generally, the higher the ICC ratio, the more likely that IA was part of a major disaster declaration.

FEMA emphasized that the proposed approach would not be a hard "threshold," nor did it anticipate incorporating it into its regulations because one single factor would not determine each decision on recommendations to the President.

A hard "threshold" was a recommendation of some FEMA stakeholders. It is likely they wanted to see a number comparable to the per capita amount employed as a factor for PA declarations. However, in response to that request, FEMA invoked Section 320 of the Stafford Act and suggested it chose not to violate the principles of Section 320. That section forbids a denial of assistance based *solely* (emphasis added) on an

---

[14] See U.S. General Accounting Office, *Disaster Assistance: Improvement Needed in Disaster Declaration Criteria and Eligibility Assurance Procedures*, GAO-01-837, August 31, 2001, available at http://www.gao.gov/assets/240/232622.pdf.

FEMA Individual Assistance Programs: In Brief 61

"arithmetic formula." While some formulas have been routinely used by FEMA in the process of making recommendations, according to FEMA, formulas have only been one factor among several considered. Instead, FEMA indicates its desire that such formulas, when used as a part of the PDA process, would influence a state's decision on whether to make a request, but would not remove the state's discretion.

FEMA's proposal offered more detail than previously available regarding state capacity, the tools FEMA may employ to consider individual state resources, and FEMA's own approach to assessing that capacity. But the proposed ICC and other factors raise questions as to how various factors may or may not be weighted in importance when considering a governor's request.

In addition, the proposed factors appeared to relate directly to current IA programs such as Disaster Unemployment Assistance and Crisis Counseling. The proposed rule also appeared to encourage the development of state programs that might replicate or complement FEMA-style assistance to families and individuals at the state level. It could be argued that such programs at the state level could be of assistance when federal supplemental help is not contemplated. At this time there is no information on when FEMA intends to finalize this proposed rule on revising IA factors.

In: Issues in Disaster Recovery and Assistance ISBN: 978-1-53616-308-7
Editor: Donatien Moïse © 2019 Nova Science Publishers, Inc.

Chapter 3

FEDERAL DISASTER ASSISTANCE: INDIVIDUAL ASSISTANCE REQUESTS OFTEN GRANTED, BUT FEMA COULD BETTER DOCUMENT FACTORS CONSIDERED[*]

United States Government Accountability Office

ABBREVIATIONS

DHS	Department of Homeland Security
FEMA	Federal Emergency Management Agency
IA	Individual Assistance
RVAR	Regional Administrator's Validation and Recommendation

---

[*] This is an edited, reformatted and augmented version of the United States Government Accountability Office Report to Congressional Requesters, Publication No. GAO-18-366, dated May 2018.

| Stafford Act | Robert T. Stafford Disaster Relief and Emergency Assistance Act |

WHY GAO DID THIS STUDY

FEMA's IA program provides help to individuals to meet their immediate needs after a disaster, such as shelter and medical expenses. When a state, U.S. territory, or tribe requests IA assistance through a federal disaster declaration, FEMA evaluates the request against regulatory factors, such as concentration of damages, and provides a recommendation to the President, who makes a final declaration decision.

GAO was asked to review FEMA's IA declaration process. This chapter examines (1) the number of IA declaration requests received, declared, and denied, and IA actual obligations from calendar years 2008 through 2016, (2) the extent to which FEMA accounts for the regulatory factors when evaluating IA requests, and (3) any challenges FEMA regions and select states reported on the declaration process and factors and any FEMA actions to revise them. GAO reviewed FEMA's policies, IA declaration requests and obligation data, and FEMA's RVARs from July 2012 through December 2016, the most recent years for which data were available. GAO also reviewed proposed rulemaking comments and interviewed FEMA officials from all 10 regions and 11 state emergency management offices selected based on declaration requests and other factors.

WHAT GAO RECOMMENDS

GAO recommends that FEMA evaluate why regions are not completing the RVARs for each element of the current IA regulatory factors and take corrective steps, if necessary. DHS concurred with the recommendation.

WHAT GAO FOUND

From calendar years 2008 through 2016, the Department of Homeland Security's (DHS) Federal Emergency and Management Agency (FEMA) received 294 Individual Assistance (IA) declaration requests from states, U.S. territories, and tribes to help individuals meet their immediate needs after a disaster. Of these, the President declared 168 and denied 126 requests. Across the various types of IA declaration requests, severe storms (190) were the most common disaster type and drought (1) was among the least common. FEMA obligated about $8.6 billion in IA for disaster declarations during this period.

GAO found that FEMA regions did not consistently obtain and document information on all elements of established IA regulatory factors when making IA recommendations to headquarters. Following a declaration request, a FEMA region is to prepare a Regional Administrator's Validation and Recommendation (RVAR)—a document designed to include data on each of the six IA regulatory factors for each declaration request as well as the regional administrator's recommendation. GAO reviewed all 81 RVARs from July 2012—the date FEMA began using the new RVAR template—through December 2016. GAO found that regions did not consistently obtain and document information for the elements required under the six regulatory factors (see table). For example, only 44 of the 81 RVARs documented all elements under the concentration of damage factor. By evaluating why regions are not completing all elements of each current IA regulatory factor, FEMA could identify whether any corrective steps are needed.

Officials from the 10 FEMA regions and 11 states GAO interviewed, reported positive relationships with each other, but also cited various challenges with the IA declaration process and regulatory factors. For example, these officials told GAO that there are no established minimum thresholds for IA, making final determinations more subjective and the rationale behind denials unclear. However, as required by the Sandy Recovery Improvement Act of 2013, FEMA has taken steps to revise the IA factors by issuing a notice of proposed rulemaking. According to

United States Government Accountability Office

FEMA, the proposed rule aims to provide more objective criteria, clarify the threshold for eligibility, and speed up the IA declaration process. As of April 2018, the proposed rule was still under consideration. According to FEMA officials, they plan to finalize the rule in late 2018; therefore, it is too early to know the extent to which it will address these challenges.

Analysis of 81 Regional Administrator's Validation and Recommendations by Element for Each Individual Assistance (IA) Regulatory Factor Documented from July 2012 through December 2016

IA Regulatory Factor	All elements documented	Some elements documented	No elements documented
Concentration of damages	44	37	0
Trauma	30	51	0
Special populations	72	8	1
Voluntary agency assistance	76	5	0
Insurance coverage	5	73	3
Average amount of IA by state	11	66	4

Source: GAO analysis based of Federal Emergency Management Agency's Regional Administrator's Validation and Recommendations. I GAO-18-366.

May 31, 2018

The Honorable Kevin McCarthy
Majority Leader
House of Representatives

The Honorable Dianne Feinstein
United States Senate

Natural disasters in 2017, such as the California wild fires and the Atlantic hurricane season, affected approximately 25.8 million people in the United States—nearly 8 percent of the U.S. population. As a result of these disasters, 4.7 million individuals applied for assistance in 2017 from the Federal Emergency Management Agency (FEMA) through its

Individual Assistance (IA) program, which provides help to individuals and families to meet their immediate needs, shelter, and medical needs in the wake of a disaster. FEMA provided more than $2 billion in IA funds in response to these disasters. FEMA is a component of the Department of Homeland Security (DHS) that leads the federal effort to prepare, respond to, and help recover from disasters, both natural and man-made. Following a major disaster declaration by the President, FEMA may provide three principal forms of assistance.[1] These include IA; Public Assistance, which addresses repairs to communities' and states' infrastructure; and Hazard Mitigation Assistance, which provides funding for projects a state submits to reduce the threat of future damage.

To obtain federal disaster assistance, a state or tribe must request a disaster declaration through FEMA (IA declaration), and then FEMA determines whether to make a recommendation to the President to declare a major disaster.[2] In reviewing the state or tribe's IA disaster declaration request, FEMA is to consider six primary factors, established in regulation in 1999, to determine the severity, magnitude, and impact of a disaster event.[3] These IA regulatory factors include: (1) concentration of damages (e.g., homes destroyed); (2) trauma (e.g., injuries and death); (3) special populations (e.g., elderly and disabled); (4) voluntary agency assistance; (5) insurance coverage; and (6) average amount of individual assistance by state. After Hurricane Sandy, Congress passed the Sandy Recovery

---

[1] In response to a request from a governor of a state or the chief executive of an affected Indian tribal government, the President may declare that a major disaster or emergency exists. 42 U.S.C. §§ 5170, 5191. If the President declares an emergency, rather than a major disaster, the federal government may provide immediate and short-term assistance that is necessary to save lives, protect property and public health and safety, or lessen or avert the threat of a catastrophe, among other things. 42 U.S.C. § 5192. Federal assistance may not exceed $5 million under an emergency declaration unless continued emergency assistance is immediately required; there is a continuing and immediate risk to lives, property, public health or safety; and necessary assistance will not otherwise be provided on a timely basis. 42 U.S.C. § 5193. Additionally, upon the request of a governor, the President may issue a fire assistance declaration that provides financial and other assistance to supplement state and local firefighting resources for fires that threaten destruction that might warrant a major disaster declaration. 44 C.F.R. § 204.21. Hereafter in this report, major disaster declarations are referred to as disaster declarations.

[2] Throughout this report, and in accordance with the Stafford Act, "state" means any state of the United States, the District of Columbia, Puerto Rico, the Virgin Islands, Guam, American Samoa, and the Commonwealth of the Northern Mariana Islands. See 42 U.S.C. § 5122(4).

[3] 44 C.F.R. § 206.48(b).

68   *United States Government Accountability Office*

Improvement Act of 2013 to improve certain aspects of disaster assistance, and the act directed FEMA to review, update, and revise the IA factors in order to provide more objective criteria, clarify the threshold for eligibility, and speed the IA declaration process.[4]

You asked us to review the IA declaration process and regulatory factors along with FEMA's proposed changes to the factors. This chapter examines (1) the number of IA declaration requests received, declared, and denied, including the types of disasters and related obligations for IA major disaster declarations from calendar years 2008 through 2016; (2) the extent to which FEMA accounts for the six IA factors when evaluating state and tribal IA declaration requests; and (3) what challenges, if any, FEMA regions and selected states report regarding the factors used in the IA declaration process, and what actions, if any, FEMA has taken to revise these factors.

To answer our first objective, we reviewed FEMA policies and procedures, regulations, and internal documents related to the IA disaster declaration process such as FEMA's guidance on the process, manual on damage assessment, and fact sheets on IA programs. We obtained and analyzed data from FEMA's systems for disaster declaration requests made by states and tribal entities; and IA actual obligations for declarations made from calendar years 2008 through 2016, the most recent data available at the beginning of our review.[5] To assess the reliability of these data, we reviewed the data and discussed data quality control procedures with FEMA officials. We determined that the data we used from these systems were sufficiently reliable for the purposes of this chapter.

To answer our second objective, we reviewed relevant laws, FEMA policies and procedures, regulations, guides, memoranda, internal documents, and other documents related to the IA disaster declarations process, including the current IA factors established in regulation. We reviewed and analyzed the completeness of all 81 FEMA nonemergency and nonexpedited Regional Administrator's Validation and

---

[4] Pub. L. No. 113-2, § 1109, 127 Stat. 4, 47 (2013).

[5] We requested data on disaster declaration requests, including IA requests, from FEMA's National Emergency Management Information System and IA actual obligations from FEMA's Integrated Financial Management Information System.

Federal Disaster Assistance 69

Recommendations (RVAR) prepared from July 2012 through December 2016, the most recent years for which data were available at the time of our review.[6] We excluded emergency and expedited disaster declaration requests because such requests are not required to include all the information related to the regulatory factors. We did not include RVARs prepared prior to July 2012, after FEMA updated its RVAR guidance and training and began recommending the use of a new RVAR template in order to help ensure consistency across regions.[7] We also obtained a copy of the RVAR template, which we determined contained 28 elements, most of which pertain to the six IA factors identified in the regulation.[8] We used these elements to analyze the 81 RVARs to determine if information on the IA regulatory factors were obtained and documented. We interviewed officials from FEMA headquarters and all 10 FEMA regions on their consideration of the IA regulatory factors and development of the RVARs. We also compared the 81 RVARs to *Standards for Internal Control in the Federal Government*.[9] To assess the reliability of the RVARs, we reviewed and analyzed the RVARs that FEMA officials provided and discussed quality control procedures with them. We determined that the information we used from the RVARs was sufficiently reliable for the purposes of this chapter.

---

[6] RVAR is a document submitted by the FEMA regional administrator to FEMA headquarters, comprising a state's or tribe's disaster declaration request, information on the IA regulatory factors compiled through regional assessments of the request, and the regional administrator's validation and recommendation regarding the request.

[7] FEMA staff began using the RVAR template for events occurring June 13, 2012, or later. As such, we limited our review of the RVARs from July 1, 2012 in order to analyze consistent data.

[8] See 44 C.F.R. 206.48(b). The elements identified in FEMA's RVAR template include: (1) total estimated federal obligation, (2) recent disasters in the same area in the past/within 12 months, (3) damage to critical facilities, (4) hazard mitigation measures, (5) impact and frequency of prior disasters, (6) level of volunteer assistance, (7) other federal assistance available, (8) trauma-death, (9) trauma-injury, (10) trauma-power outage, (11) trauma-disruption of community functions/services, (12) damage concentration, (13) homes destroyed, (14) homes with major damage, (15) homes with minor damage, (16) homes affected, (17) home ownership, (18) insurance, (19) flood insurance, (20) low income, (21) median household income, (22) poverty, (23) disabled, (24) elderly, (25) Small Business Administration loans, (26) resident loans, (27) Small Business Administration number of loans, and (28) resident number of loans.

[9] GAO, *Standards for Internal Control in the Federal Government*, GAO-14-704G (Washington, D.C.: September 2014).

To answer our third objective, we reviewed relevant laws and FEMA policies and procedures, notices, regulations, and guidance related to the IA disaster declaration process and FEMA's proposed rulemaking, which proposes revisions to the current six IA regulatory factors. We interviewed officials from all 10 FEMA regions, FEMA headquarters officials, and state-level emergency management officials in 11 states to discuss their perspectives on the IA declaration process and current regulatory factors, any associated challenges, and their perspectives on the proposed changes to the current IA regulatory factors. We selected the 11 states based on those that have undergone multiple IA requests from calendar years 2008 through 2016 and those that have experienced various types of disasters in different regions of the country. We selected California, Illinois, Indiana, Kentucky, Mississippi, Missouri, New York, Ohio, Oklahoma, South Dakota, and Virginia. We made three site visits to obtain more detailed information on the IA declaration processes. We visited FEMA regions IX and V, and spoke with regional FEMA officials, as well as with emergency management officials for the states of California and Illinois. We also visited region IV and we spoke with FEMA officials for that region. We chose these locations to obtain examples and experiences across a wide variety of disaster types and regions. The information gathered during these site visits is not generalizable to other states or regions, but the details provide insights regarding FEMA's and states' management of these processes. We also reviewed and summarized the 14 public comments states submitted on FEMA's proposed rulemaking that proposes changes to the current IA regulatory factors.

We conducted this performance audit from January 2017 to May 2018 in accordance with generally accepted government auditing standards. Those standards require that we plan and perform the audit to obtain sufficient, appropriate evidence to provide a reasonable basis for our finding and conclusions based on our audit objectives. We believe that the evidence obtained provides a reasonable basis for our findings and conclusions based on our audit objectives.

BACKGROUND

The Robert T. Stafford Disaster Relief and Emergency Assistance Act (Stafford Act), as amended, establishes the process for states or tribal entities to request a presidential disaster declaration.[10] The act also generally defines the federal government's role during the response and recovery after a disaster and establishes the programs and process through which the federal government provides disaster assistance to state, local governments, tribal entities and individuals.[11] In addition to its central role in recommending to the President whether to declare a disaster, FEMA has primary responsibility for coordinating the federal response when a disaster is declared as well as recovery, which typically consists of providing grants to assist state and tribal entities to alleviate the damage resulting from such disasters. Once a disaster is declared, FEMA provides assistance through the IA, Public Assistance, and Hazard Mitigation Assistance programs. For instance, some declarations may provide grants only for IA and others only for Public Assistance. Hazard Mitigation Assistance grants, on the other hand, are available for all declarations if the affected area has a FEMA-approved Hazard Mitigation plan. The process for requesting assistance is the same for the three types of assistance.

Disaster Declaration Process

Under the Stafford Act, states' governors or tribal chief executives may request federal assistance, if state and tribal resources are overwhelmed after a disaster.[12] As part of the request to the President, a governor or tribal chief executive must affirm that the state or tribe has implemented an emergency plan and that the situation is of such severity and magnitude that effective response is beyond the capabilities of the state or tribal entity, among other things. After a state or tribe submits a request

---

[10] Pub. L. No. 93-288, 88 Stat. 143 (codified as amended at 42 U.S.C. § 5121 et seq.).
[11] See generally 42 U.S.C. § 5121 et seq.
[12] See 42 U.S.C. § 5170.

for disaster declaration through FEMA's regional office, the regional office is to evaluate the request and make a regional recommendation through the RVAR, which is submitted to FEMA headquarters for further review. The FEMA administrator then is to review the state's or tribe's request and the RVAR, and recommend to the President whether a disaster declaration is warranted. Figure 1 shows the process for a disaster declaration from the time a disaster occurs until the President approves or denies a declaration request.

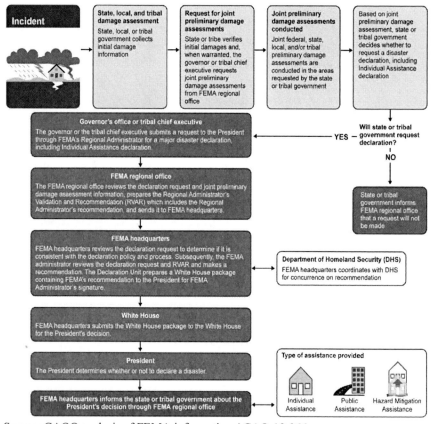

Source: GAQO analysis of FEMA information. | GAO-18-366.

Figure 1. Federal Emergency Management Agency's (FEMA) Major Disaster Declaration Process.

Federal Disaster Assistance 73

Five Programs Available under Individual Assistance

The IA program provides financial and direct assistance to disaster victims for expenses and needs that cannot be met through other means, such as insurance. The IA comprises five different programs as shown below. When states or tribal entities request disaster declarations, they may request assistance under any or all of the five programs. Likewise, when the President makes a disaster declaration, the declaration may authorize IA which may also include any or all of the five IA programs.

1. *Individuals and Households Program* provides assistance to eligible disaster survivors with necessary expenses and serious needs which they are unable to meet through other means, such as insurance.[13] According to FEMA headquarter officials, direct assistance is provided to individuals to meet housing needs.

2. *Crisis Counseling Program* assists individuals and communities by providing community-based outreach and psycho-educational services.[14]

3. *Disaster Legal Services* provides assistance through an agreement with the Young Lawyers Division of the American Bar Association for free legal help to survivors who are unable to secure legal services adequate to meet their disaster-related needs.[15]

4. *Disaster Case Management Program* involves a partnership between a FEMA disaster case manager and a survivor to develop and carry out a Disaster Recovery Plan.[16]

5. *Disaster Unemployment Assistance* provides unemployment benefits and reemployment services to individuals who have become unemployed as a result of a major disaster and who are not eligible for regular state unemployment insurance.[17]

---

[13] See 42 U.S.C. § 5174; 44 C.F.R. §§ 206.110-206.120.
[14] See 42 U.S.C. § 5183; 44 C.F.R. § 206.171.
[15] See 42 U.S.C. § 5182; 44 C.F.R. § 206.164.
[16] See 42 U.S.C. § 5189d.
[17] See 42 U.S.C. §5177; 44 C.F.R. § 206.141.

The Six IA Regulatory Factors Used to Assess IA Declaration Requests

In accordance with its responsibilities under the Stafford Act, FEMA issued a regulation in 1999 that outlines the six factors regional and headquarters officials are to consider when assessing requests for a disaster declaration and when developing a recommendation to the President for a federal disaster declaration.[18] The regulation states that FEMA considers the six factors not only to evaluate the need for IA but also to measure the severity, magnitude, and impact of the disaster. The state or tribe provides information on these factors when submitting its disaster declaration request. The six factors for IA include the following:

1. *Concentration of Damages*—characterizes the density of the damage in individual communities. The regulation states that highly concentrated damages "generally indicate a greater need for federal assistance than widespread and scattered damages throughout a state." For example, concentration of damage data includes the numbers of homes destroyed, homes with major or minor damages, and homes affected.
2. *Trauma*—the regulation provides conditions that might cause trauma including large numbers of injuries and deaths, large-scale disruption of normal community functions, and emergency needs such as extended loss of power or water.
3. *Special Populations*—FEMA considers the impact of the disaster on special populations, such as low-income populations, the elderly, or the unemployed.

---

[18] 44 C.F.R. § 206.48(b); see also 42 U.S.C. § 5170. Consistent with 42 U.S.C. § 5170, as amended by the Sandy Recovery Improvement Act of 2013, the Governor of an affected state or the Chief Executive of an affected Indian tribal government may submit a request for a declaration. See Pub. L. No. 113-2, § 1110, 127 Stat. 4, 48 (codified as amended at 42 U.S.C. §§ 5170, 5191, 5122).

Federal Disaster Assistance

4. *Voluntary Agency Assistance*—involves the availability and capabilities of voluntary, faith, and community-based organizations, and state and local programs to help meet both the emergency and recovery needs of individuals affected by disasters.

5. *Insurance Coverage*—addresses the level of insurance coverage among those affected by disasters.[19] Because disaster assistance cannot duplicate insurance coverage, as recognized in the regulation, if a disaster occurred where almost all of the damaged dwellings were fully insured for the damage that was sustained, FEMA could conclude that a disaster declaration by the President was not necessary in accordance with this factor.[20]

6. *Average Amount of Individual Assistance by State*—according to the regulation, there is no set threshold for recommending IA. However, it states that the averages, depicted in table 1, may prove useful to states and voluntary agencies as they develop plans and programs to meet the needs of disaster victims. The inference is that these averages generally indicate the amount of damages that could be expected for a state based on its size (small, medium, and large). The averages contained within the regulation and depicted in table 1 are based on disasters that occurred between July of 1994 and July of 1999.[21]

---

[19] By law, federal disaster assistance cannot duplicate insurance coverage. See 42 U.S.C. § 5155; 44 C.F.R. § 206.48(b)(5).

[20] This provision does not necessarily result in delayed assistance. FEMA is able to provide help to individuals and households that have disaster damages but are waiting on insurance or other assistance for help. Those applicants can receive FEMA help as long as they agree to reimburse FEMA when they receive their other assistance. See 44 C.F.R. § 206.113(a)(3).

[21] FEMA headquarters officials stated that the table titled "Average Amount of Assistance per Disaster by Size of State" contained within 44 C.F.R. § 206.48(b)(6) is considered outdated and is no longer relevant to the declaration process. FEMA officials indicated that to satisfy the factor of average amount of IA assistance by state they evaluate the total number of homes destroyed and suffering major damage (as well as the accessibility and habitability of the dwellings and the community).

76 *United States Government Accountability Office*

Table 1. Average Amount of Individual Assistance per Disaster by Size of State Contained in 44 C.F.R. § 206.48(b)(6) (July 1994 to July 1999)

	Small states (under 2 million population)	Medium states (2-10 million population)	Large states (over 10 million population)
Average population (1990 census data)	1,000,057	4,713,548	15,522,791
Number of Disaster Housing applications approved	1,507	2,747	4,679
Number of homes estimated major damage/destroyed	173	582	801
Dollar amount of housing assistance ($)	2.8 million	4.6 million	9.5 million
Number of Individual and Family Grant applications approved	495	1,377	2,071
Dollar amount of Individual and Family Grant assistance ($)	1.1 million	2.9 million	4.6 million
Disaster Housing/Individual and Family Grant combined assistance ($)	3.9 million	7.5 million	14.1 million

Source: 44 C.F.R. § 206.48(b)(6). I GAO-18-366.

THE PRESIDENT DECLARED 57 PERCENT OF ALL IA REQUESTS FROM 2008 THROUGH 2016, WITH TOTAL OBLIGATIONS OF APPROXIMATELY $8.6 BILLION

The Number of IA Declarations Varied by Region and Severe Storms Were the Most Frequent Disaster Type

The President declared 57 percent of all IA declaration requests from calendars years 2008 through 2016, with total IA obligations of approximately $8.6 billion. FEMA received 294 IA declaration requests from calendar years 2008 through 2016. Of these, the President declared 168 requests (57 percent), and 51 percent of these declarations were from Regions IV and VI, as shown in table 2.[22]

---

[22] Information on the composition of the FEMA regions by state or tribe related to the IA declaration requests is found in appendix I of this report.

Federal Disaster Assistance

Table 2. Information on Individual Assistance (IA) Requests and Declarations by Federal Emergency Management Agency (FEMA) Region from Calendar Years 2008 through 2016

Region	IA Requested	IA Declared	Percentage Declared (%)
I	15	13	87
II	18	10	56
III	26	13	50
IV	69	53	77
V	32	16	50
VI	59	32	54
VII	25	13	52
VIII	18	8	44
IX	18	6	33
X	14	4	29
Total	294	168	57

Source: GAO analysis of FEMA data. I GAO-18-366.

Table 3. Number of Individual Assistance (IA) Declarations Denied by Federal Emergency Management Agency (FEMA) Region from Calendar Years 2008 through 2016

Region	IA Requested	IA Denied	Percentage Denied (%)
X	14	10	71
IX	18	12	67
VIII	18	10	56
III	26	13	50
V	32	16	50
VII	25	12	48
VI	59	27	46
II	18	8	44
IV	69	16	23
I	15	2	13
Total	294	126	43

Source: GAO analysis of FEMA data. I GAO-18-366.

Additionally, of the 126 IA declaration requests denied by the President, Regions X and IX had the highest percentage of denials, at 71 percent (10 out of 14) and 67 percent (12 out of 18), respectively, and Region I had the lowest percentage of denials at 13 percent (2 out of 15),

78 *United States Government Accountability Office*

as shown in table 3. See appendix I for the number of IA declarations requested, declared, and denied by states and tribes from each FEMA region for disaster declarations requested from calendar years 2008 through 2016.

Table 4. Information on Individual Assistance (IA) Declarations Requests by Type of Disaster from Calendar Years 2008 through 2016[a]

Disaster Types	IA Requested	IA Declared	IA Denied
Severe storm	190	113	77
Flooding	149	92	57
Tornado	117	68	49
Straight-line wind	60	35	25
Hurricane	27	24	3
Wildfire	16	6	10
Mudslide	15	11	4
Tropical storm	14	9	5
Winter storm	12	2	10
Landslide	12	7	5
Earthquake	5	2	3
Snow storm	3	0	3
Typhoon	2	1	1
Ice jam	2	1	1
Tsunami	2	1	1
Explosion	2	0	2
Fishery closure	1	0	1
Contaminated water	1	0	1
Drought	1	0	1

Source: GAO analysis of Federal Emergency Management Agency data. I GAO-18-366.

[a] Each disaster declaration request may have more than one type of disaster. As such, the table does not include a row for totals.

According to a FEMA headquarters official, when a disaster declaration is denied, FEMA sends a denial letter to states or tribes based on the review of all the information available. The letter generally states that the damage was not of such severity and magnitude as to be beyond the capabilities of the state, affected local governments, and voluntary agencies, and accordingly the supplemental federal assistance is not necessary. Of the emergency management officials we interviewed in 11

Federal Disaster Assistance

states, officials in five states reported that FEMA provided a rationale behind the denial, while officials in three states reported that no rationale was provided.[23]

Among the various types of disasters for which IA declaration requests were received, severe storms, flooding, and tornados accounted for the highest number of IA requests, with drought, fishery closure, and contaminated water being the least common, as shown in table 4.

FEMA IA Obligations Varied by Region and State

Table 5. Actual Obligations of Individual Assistance by State for Disaster Declarations Made from Calendar Years 2008 through 2016[a]

State	Region	Total ($)
Louisiana	VI	2,003,244,709
New York	II	1,309,869,042
Texas	VI	1,149,629,671
New Jersey	II	656,334,017
Illinois	V	621,800,788
Iowa	VII	210,526,142
North Dakota	VIII	200,890,261
Tennessee	IV	199,447,951
North Carolina	IV	174,448,712
Pennsylvania	III	165,486,633
Michigan	V	148,308,502
South Carolina	IV	147,945,615
Alabama	IV	137,509,405
Florida	IV	121,961,283
Missouri	VII	116,280,805
Wisconsin	V	115,849,685
Indiana	V	104,041,429
West Virginia	III	100,322,247
Kentucky	IV	95,077,277
Mississippi	IV	80,929,297
Colorado	VIII	79,164,491
Georgia	IV	77,397,631

---

[23] The other 3 state emergency officials did not provide any comments on the rationale behind denied IA declaration requests in their state.

Table 5. (Continued)

State	Region	Total ($)
Massachusetts	I	70,452,076
Puerto Rico	II	58,237,566
Oklahoma	VI	57,607,547
South Dakota	VIII	55,503,363
Arkansas	VI	54,859,658
California	IX	46,917,730
Rhode Island	I	38,831,216
Mariana Islands	IX	34,281,445
American Samoa	IX	32,242,550
Connecticut	I	30,878,096
Vermont	I	30,362,291
Virginia	III	26,530,586
Washington	X	13,450,309
Alaska	X	11,386,310
Montana	VIII	6,634,838
Nebraska	VII	6,160,220
Minnesota	V	5,532,091
Hawaii	IX	4,082,910
Wyoming	VIII	3,029,312
Maryland	III	2,649,629
Nevada	IX	1,893,834
New Hampshire	I	1,260,630
Maine	I	1,229,976
Virgin Islands	II	2,136
Total		8,610,481,911

Source: GAO analysis of Federal Emergency Management Agency data. I GAO-18-366.

[a] Table includes actual obligations made from the time of the declaration to March 2017. In accordance with the Stafford Act, "state" means any state of the United States, the District of Columbia, Puerto Rico, the Virgin Islands, Guam, American Samoa, and the Commonwealth of the Northern Mariana Islands. See 42 U.S.C. § 5122(4).

Federal Disaster Assistance

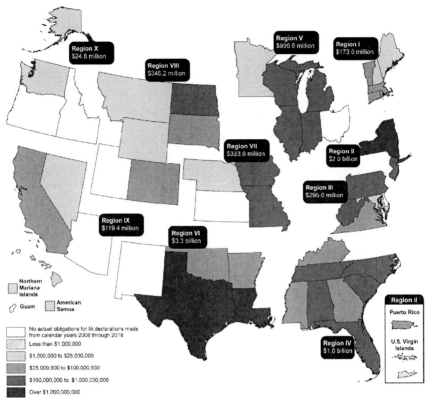

Source: GAO analysis of FEMA data, Map Resources (map). | GAO-18-366.

[a] Figure includes IA actual obligations made from the time of the declaration to March 2017. In accordance with the Stafford Act, "state" means any state of the United States, the District of Columbia, Puerto Rico, the Virgin Islands, Guam, American Samoa, and the Commonwealth of the Northern Mariana Islands. See 42 U.S.C. § 5122(4).

Figure 2. Federal Emergency Management Agency's (FEMA) Individual Assistance (IA) Actual Obligations by Regions for Disaster Declarations Made from Calendar Years 2008 through 2016.[a]

FEMA obligated a total of approximately $8.6 billion in IA for disaster declarations made from calendar years 2008 through 2016. These actual obligations were provided to 46 states and they ranged from less than $1 million to more than $1 billion as shown in figure 2.[24]

See appendix II for FEMA's IA actual obligations by state and type of disasters for disaster declarations made from calendars years 2008 through 2016. Additionally, actual obligations for IA declarations made from calendar years 2008 through 2016 varied greatly by FEMA region, as also shown in figure 3. For example, FEMA Region VI had the highest obligations at around $3.3 billion. Region X had the lowest obligations at $24.8 million. As shown in table 5, the amount of obligations for disasters declarations also varied greatly by state. For example, Louisiana had the highest obligations at approximately $2 billion, followed by New York and Texas at about $1.3 billion and $1.1 billion, respectively. The state with the lowest obligations was the U.S. Virgin Islands at about $2,100.

FEMA REGIONS VARIED IN HOW THEY CONSIDERED IA REGULATORY FACTORS AND DID NOT CONSISTENTLY OBTAIN AND DOCUMENT INFORMATION ON ALL ELEMENTS OF THESE FACTORS

FEMA Regions Varied in Their Consideration of the IA Regulatory Factors Based on Disaster Circumstances

Six of FEMA's 10 regional offices reported using all six regulatory factors when evaluating states' or tribes' IA declaration requests. Officials from the other 4 regions reported using five of the six factors, with the exception being the average amount of individual assistance by state factor. These officials noted that they do not use this factor because FEMA

---

[24] Throughout this report, actual obligations refer to funds obligated by FEMA and are not adjusted for inflation. Also, these actual obligations were made from the time of the declaration to March 2017.

considers the factor to be outdated or they consider all of the factors holistically.[25]

Officials from FEMA's regional offices also generally reported that the extent to which they consider the six IA regulatory factors equally in all cases varies, depending on the circumstances of the related disaster. Specifically, officials from 7 of the 10 regions stated that they use the regulatory factors on a case-by-case basis as certain factors are more relevant than others based on the disaster. For example, if a tornado hits a rural community and completely destroys all properties within the community with no death or injury, then the regulatory factor for trauma may not be as applicable, while the concentration of damages regulatory factor would have greater relevance. On the other hand, if a tornado hits the center of a town resulting in damages with death and injuries, then the trauma regulatory factor would become more important to consider.

Additionally, officials in 3 of the 10 regions reported that in addition to the six regulatory factors, they also take into account institutional knowledge and staff experience when evaluating the regulatory factors. For example, officials in one region stated that their staff have more than 10 years of IA declaration experience, and as such, they are familiar with the extent of the information needed and collect the information accordingly.

FEMA Regions Did Not Consistently Obtain and Document Information on All Elements of the IA Regulatory Factors in RVARs

Based on our analysis of RVARs from July 2012 through December 2016 used to recommend approving or denying IA requests, FEMA regional offices did not consistently obtain and document information on

---

[25] FEMA headquarters officials indicated that the table titled "Average Amount of Assistance per Disaster by Size of State" contained within contained within 44 C.F.R. § 206.48(b)(6) is considered outdated and is no longer relevant to the declaration process. FEMA officials stated that they evaluate the total number of homes destroyed and suffering major damage (as well as the accessibility and habitability of the dwellings and the community) to satisfy this factor.

all elements of the IA regulatory factors. As described earlier, FEMA regions are to use the RVAR to document information on the IA factors and to recommend to the FEMA administrator whether a disaster should be declared.

According to FEMA headquarters officials, FEMA developed the RVAR template in June 2012 to help ensure consistency across regions when making recommendations to headquarters on IA declaration requests. Officials stated that prior to the template, information on the six factors was mainly provided in narrative format. The new template listed the various elements found within each of the six regulatory factors, guiding the regional offices to provide information based on those elements. For example, instead of providing a general narrative on the trauma factor, the new template listed the elements to be provided under trauma, such as the number of injuries and deaths, as well as information on power outages and disruption of other community functions and services. Also, instead of summarizing the concentration of damages factor, the template allowed regional offices to categorize the damage concentration as low, medium, high, or extreme. Furthermore, the template also provided a uniform format to present quantitative information such as the number of homes destroyed; whether home damages are major or minor; the number of homes affected; and level of home ownership. See appendix III for a sample RVAR template.

We analyzed 81 RVARs developed by the 10 FEMA regions from July 2012 through December 2016 and found that regions did not consistently obtain and document information on all elements related to each of the six regulatory factors in their RVARs.[26] As shown in table 6, all 81 RVARs

---

[26] As mentioned earlier, the RVAR template contains 28 elements, most of which pertain to the six IA factors based on 44 C.F.R. § 206.48(b). For example, the concentration of damages factor contains 6 elements (e.g., homes destroyed, homes with major damage, homes with minor damage, homes affected, damage to critical facilities, and damage concentration); trauma (e.g., injuries, death, power outages, and disruption of community functions/services) and special populations (e.g., low-income, poverty, disabled, and elderly) contain 4 elements; insurance has 3 elements (e.g., home ownership, insurance, and flood insurance [if applicable]); and voluntary agency assistance (e.g., hazard mitigation measures and level of volunteer assistance) and average amount of IA by state (e.g., recent disasters in the past 12 months and impact and frequency of prior disasters) have 2 elements.

had at least some elements documented but not all for each of the IA regulatory factors. For example, for the IA concentration of damages regulatory factor, the six elements to be addressed include the number of homes destroyed, damaged or affected, damage concentration, and damage to critical facilities. While 44 of the 81 RVARs documented all of the six elements, 37 documented some but not all of the elements. Similarly, for the trauma regulatory factor, the four elements to be addressed include injuries, death, power outages, and disruption of community functions. While 30 of the 81 RVARs documented all of the four elements, 51 documented some but not all of the elements. For the insurance coverage factor, while five RVARs documented all of the elements, 73 RVARs documented some but not all of the elements. Elements under this factor include home ownership, insurance, and flood insurance, when applicable.[27] None of the six regulatory factors were fully documented across all RVARs. See appendix IV for detailed information on the extent to which all of the elements of the six regulatory factors were documented in the RVARs from July 2012 through December 2016.

Table 6. Analysis of 81 Regional Administrator's Validation and Recommendations (RVAR) by Element Pertaining to Each Individual Assistance (IA) Regulatory Factor Documented from July 2012 through December 2016

IA regulatory factor	All elements documented	Some elements documented	No elements documented
Concentration of Damages	44	37	0
Trauma	30	51	0
Special populations	72	8	1
Voluntary agency assistance	76	5	0
Insurance Coverage	5	73	3
Average amount of IA by state	11	66	4

Source: GAO analysis of Federal Emergency Management Agency's RVAR. I GAO-18-366.

FEMA headquarters officials acknowledged that information related to all the elements for each of the IA regulatory factors were missing from the

---

[27] The flood insurance element is only applicable when the disaster type is flood related.

RVARs. They stated that they had not collected all information on all factors because one factor may have more weight than another based on the specific incident that has occurred. However, they also indicated that they do not fully know and have not evaluated all of the reasons why a region may have omitted information on an element of a factor. FEMA headquarters officials agreed that having complete information on all elements of the regulatory factors in the RVARs would assist in their recommendation process. *Standards for Internal Control in the Federal Government* suggest that agencies should establish and operate monitoring activities to ensure that internal controls—such as the documentation of all of the elements of the IA regulatory factors FEMA regions considered—are effective, and to take corrective actions as appropriate.[28] Because it is unclear why regions are not completely documenting all elements related to the current six regulatory factors, such an evaluation could help FEMA identify whether any corrective steps are needed. Doing so could help FEMA ensure it is achieving its stated goals in providing consistency in the evaluation process and in the types of factors it considers.

FEMA AND STATES REPORTED CHALLENGES IN THE IA DECLARATION PROCESS, AND FEMA IS REVISING THE REGULATORY FACTORS USED TO ASSESS DECLARATION REQUESTS

FEMA and State Officials Reported Both Positive Relationships and Some Challenges in the IA Declaration Process

Officials we interviewed in 9 of the 10 FEMA regions and state emergency management offices in all 11 states reported the positive relationship they maintain with each other as a strength in the IA

---

[28] GAO, *Standards for Internal Control in the Federal Government*, GAO-14-704G (Washington, D.C.: September 2014).

Federal Disaster Assistance

87

declaration process.[29] For example, both FEMA regional officials and state emergency management officials stated that they have a good working relationship and are in regular communication via telephone or in-person meetings with each other. Also, state emergency management officials we spoke to stated that whenever they are in need of assistance, they know they can reach out to FEMA regional officials for assistance. However, FEMA regional and state emergency management officials we spoke to also reported various challenges with the process. These include the subjective nature of the IA regulatory factors given the lack of eligibility thresholds, the lack of transparency in the decision-making process, and difficulty gathering information on IA regulatory factors.

Subjective Nature of the IA Factors and Lack of Eligibility Thresholds

Officials from 9 of 10 FEMA regions stated the subjective nature of the IA program is a challenge; and officials in 6 of the 10 regions also said they found the lack of eligibility thresholds a challenge. An official in one region stated that unlike FEMA's Public Assistance program, which has minimum thresholds for eligibility, it is unclear when states should apply for IA funds. Under the Public Assistance program, for example, for states or tribes to qualify for assistance, they must demonstrate that they have sustained a minimum of $1 million in damages and the impact of damages must amount to $1.00 per capita in the state.[30] An official in another region explained that although the subjectivity of the IA factors provides flexibility in determining the type of IA program needed, having some quantifiable criteria could help officials explain to states why their requests were denied or approved.

---

[29] The remaining one regional office did not make any comment on the FEMA and state relationship. We conducted interviews with all 10 FEMA regions and 11 states, including California, Illinois, Indiana, Kentucky, Mississippi, Missouri, New York, Ohio, Oklahoma, South Dakota, and Virginia.

[30] See 44 C.F.R. § 206.48(a).

88 *United States Government Accountability Office*

**Table 7. Example of Regional Administrator's Validation
and Recommendations Information on the Individual
Assistance Regulatory Factors for Four States Affected
by the Same Disaster in 2012**

Region	V	V	V	IV
State	Ohio	Illinois	Indiana	Kentucky
Disaster period	March 2	February 29-March 2	February 29-March 3	February 29-March 3
Type of disaster declaration request	Major disaster declaration	Major disaster declaration	Major disaster declaration	Expedited disaster declaration[a]
Type of disaster	Tornadoes			
Factor/Damage concentration				
Homes destroyed	19	104	187	510
Homes with major damage	23	50	85	244
Homes with minor damage	25	139	107	Not provided
Homes affected	78	133	89	Not provided
Damage to critical facilities	Not provided	Yes	Yes	Not provided
Damage concentration	Provided	Provided	Provided	Provided
Trauma				
Injury	Not provided	100	Not provided	300
Death	3	6	13	21
Power outages	Yes	Yes	Yes	Yes
Disruption of community functions/services	Not provided	Yes	Yes	Yes
Special population				
Poverty (%)	9.3	16.90	12.6	18.40
Elderly (%)	11.8	17.90	13.75	12.90
Low-income (%)	1	1	24.7	29.30
Disabled (%)	Not provided	Not provided	Not provided	23.7
Voluntary agency assistance				
Hazard mitigation measures	Yes	Yes	Yes	Yes
Level of voluntary assistance	Provided	Provided	Provided	Provided
Insurance				
Insurance (%)	97	20	47	Not provided
Flood insurance (if applicable)[b]	Not applicable	Not applicable	Not applicable	Not provided
Home ownership (%)	99	65	93	Not provided

Region	V	V	V	IV
State	Ohio	Illinois	Indiana	Kentucky
Average amount of individual assistance by state[c]				
Recent disasters in the past 12 months	Recent disasters in the past 12 months	Recent disasters in the past 12 months	Recent disasters in the past 12 months	Recent disasters in the past 12 months
Impact and Frequency of prior disasters	Not provided	Not provided	Not provided	Yes
IA declaration request granted	No	No	Yes	Yes

Source: GAO analysis of Federal Emergency Management Agency information. I GAO-18-366.

[a] Expedited disaster declaration requests are not required to provide all the information related to the regulatory factors.

[b] Flood insurance element is only applicable when the disaster type is flood related.

[c] According to FEMA headquarters officials, the average amounts of individual assistance per disaster identified in the average amount of individual assistance by state regulatory factor, are considered outdated and they are no longer relevant to the declaration process. Instead, FEMA officials stated that they evaluate the total number of homes destroyed and suffering major damaged to satisfy this factor.

Similarly, officials we interviewed in 7 of the 11 states said they found the subjective nature of the factors with no threshold to be a challenge. A state emergency management official in one state said this subjectivity makes it difficult to determine whether or not the state should make an IA request. A state emergency management official in another state reported that the subjectivity can cause the IA declaration process to be inconsistent, and it is not always clear how or why certain declarations were approved and others were not. Further, a state emergency management official in an additional state also pointed to the subjective nature of the factors with no threshold as a reason for not being able to provide a more detailed rationale behind a declaration denial.

To illustrate this, table 7 shows how four states requested IA declarations related to the same tornado in 2012 and varied in what they reported across the six IA factors, such as the levels of damages incurred, special populations among their residents, and insurance coverage. Two of these four states—Kentucky and Indiana—received IA declarations and the other two—Ohio and Illinois—were denied.

Lack of Transparency

Another challenge reported by FEMA regional and state emergency management officials was the lack of transparency in how FEMA evaluates and provides a recommendation to the President on whether a declaration is warranted. For example, officials we interviewed in 4 of 10 regions indicated the lack of transparency as a challenge. A FEMA official in one region stated that the region would like more transparency regarding what FEMA headquarters recommends to the President and whether the President's decision aligns with FEMA's recommendation. State emergency management officials we interviewed in 10 of 11 states also reported that lack of transparency with the IA process is a challenge. For example, an emergency management official in one state said it is not clear how or if FEMA considers all of the factors. Also, an emergency management official in another state reported that it was unclear to him why his state's declaration request was denied while the requests of other states with similar incidents were declared.

Difficulty Gathering Information on IA Regulatory Factors

Officials in 4 of 10 FEMA regions reported difficulty gathering information, such as income or insurance coverage, as a challenge. An official in one region stated that it is difficult to obtain information related to IA factors from states. For example, the official said that calculating the concentration of damages is difficult absent technical guidance from FEMA headquarters, as the current guidance only accounts for the number of structure damage but not the impact of damage. Further, officials in two FEMA regions stated that states lack a dedicated IA official, making it difficult for state officials, who play multiple roles, to provide the necessary information related to the IA factors in their IA declaration request. Additionally, a state emergency management official in one state also reported that lack of staff resources in her state makes it difficult to verify all the local damage assessments prior to making a declaration request.

FEMA Is Taking Steps to Revise the IA Regulatory Factors

Pursuant to the Sandy Recovery Improvement Act of 2013, in November 2015, FEMA issued a Notice of Proposed Rulemaking to revise the six current IA regulatory factors to the following proposed factors:

- state fiscal capacity and resource availability;
- uninsured home and personal property losses;
- disaster-impacted population profile;
- impact to community infrastructure;
- casualties; and
- disaster-related unemployment.[31]

According to FEMA headquarters officials, the revisions aim to provide more objective criteria, clarify the threshold for eligibility, and speed the declaration. The officials said the proposed rule also seeks to provide additional clarity and guidance for all the established factors.[32] Table 8 shows FEMA's description of current and proposed IA factors.

FEMA received public comments from 14 states in the Federal Register during the comment period for the proposed rule and proposed guidance.[33]

---

[31] 80 Fed. Reg. 70,116 (proposed Nov. 12, 2015); see Pub. L. No. 113-2, § 1109, 127 Stat. at 47 (directing the Administrator of FEMA to, in cooperation with representatives of state, tribal, and local emergency management agencies, review, update and revise through rulemaking the individual assistance factors in order to provide more objective criteria for evaluating the need for assistance to individuals, clarify the threshold for eligibility, and speed a declaration of major disaster or emergency).

[32] In September 2016, FEMA also published and made available for comment a draft Individual Assistance Declarations Guidance, which is meant to be proposed companion guidance to accompany the proposed rule. See 81 Fed. Reg. 65,369 (published Sept. 22, 2016). In the draft guidance, FEMA describes how it plans to evaluate a governor's request for a major disaster declaration authorizing Individual Assistance, including a general description of each proposed factor, why it is important, and the data sources in the proposed factor.

[33] FEMA received 58 comments on the proposed rule of which one was withdrawn. As a result, in total, 57 comments were submitted by 24 state and county government employees representing 14 states, 5 congressional members' offices, 8 public and private organizations, and 20 individuals.

United States Government Accountability Office

Table 8. Federal Emergency Management Agency's (FEMA) Description of Current and Proposed Individual Assistance Regulatory Factors

Current Factors	Proposed Factors
Concentration of Damage	State Fiscal Capacity and Resource Availability
Trauma	Uninsured Home and Personal Property Losses
Special Populations	Disaster-Impacted Population Profile
Voluntary Agencies	Impact to Community Infrastructure
Insurance	Casualties
Average Amount of Individual Assistance by State	Disaster Related Unemployment

Source: GAO analysis of FEMA information. I GAO-18-366.

The 14 states expressed concern about the proposed factor for state fiscal capacity and resource availability, including the reliability and relevance of data sources such as total taxable resources. These states expressed concern that the data collection necessary to meet the new requirements would fall upon them, adding to the cost burden of completing an IA disaster declaration request. They also explained that the use of total taxable resources and other similar data is not an effective way to assess a state's current ability to provide resources following a disaster. Also, these states indicated that the data points such as total taxable resources and per capita personal income that would be used to evaluate state fiscal capacity are outdated and inaccurate and would be an inefficient way to evaluate a state's true fiscal capacity to respond to a disaster.

Regarding the other five proposed factors, several states in their comments raised questions about ambiguities in interpreting the factors or the feasibility and cost of gathering related data. For example, in regards to the factor on disaster impacted population, five states expressed concern that the data required for the disaster-impacted population factor would be a cost burden to the state or that the data would be inappropriate for evaluation. Additionally, two states said unemployment related to a disaster incident for the disaster-related unemployment factor would be hard to quantify in the first 30 days following a disaster. They stated that this was especially an issue given that states work to submit an IA disaster declaration request as soon as possible following a disaster.

According to the Office of Management and Budget's Office of Information and Regulatory Affairs website, the projected date for finalization of the proposed rule is September 2018; however, as of April 2018, FEMA officials stated that they were not certain whether that timeframe would be met. Until the proposed rule is finalized, we will not know the extent to which the various challenges FEMA regions and state officials raised in our interviews and in comments on the proposed rule will be addressed.

CONCLUSION

FEMA has obligated over $8.6 billion nationwide in IA from calendar years 2008 through 2016, highlighting the importance of FEMA's evaluation of states' and tribes' IA declaration requests. FEMA's regional offices evaluate the request and make a regional recommendation through the Regional Administrator's Validation and Recommendation, which documents information on all relevant IA regulatory factors. FEMA has developed the Regional Administrator's Validation and Recommendation to ensure regions consistently obtain and document the information needed by FEMA to make a disaster declaration recommendation to the President based on the IA regulatory factors. However, FEMA's regional offices do not consistently obtain and document information on all elements of the current IA regulatory factors. Because it is unclear why regions are not always documenting all of the elements related to these factors, evaluating the reasons why could help FEMA identify if any corrective steps are needed. Doing so could also help FEMA ensure it is meeting its stated goals in providing consistency in the evaluation process and in the types of factors it considers.

RECOMMENDATION FOR EXECUTIVE ACTION

We recommend that the Administrator of FEMA evaluate why regions are not completing the Regional Administrator's Validation and Recommendations for each element of the current IA regulatory factors and take corrective steps, if necessary.

AGENCY COMMENTS AND OUR EVALUATION

We provided a draft of this chapter to DHS for its review and comment. DHS provided written comments, which are summarized below and reproduced in full in appendix V. DHS concurred with the recommendation and described planned actions to address it. In addition, DHS provided written technical comments, which we incorporated into the report as appropriate.

DHS concurred with our recommendation that FEMA evaluate why regions are not completing the Regional Administrator's Validation and Recommendations for each element of the IA regulatory factors and take corrective steps, if necessary. DHS stated that a FEMA working group consisting of headquarters stakeholders will draft survey questions for FEMA region officials to identify the common reasons why an element of an IA regulatory factor may not be addressed within a RVAR. According to DHS, the working group will also analyze, assess, and present the findings of the survey responses to FEMA senior leadership, and if needed, FEMA will develop and send a memorandum to the regions with additional guidance regarding the appropriate preparation of RVARs. DHS stated that the estimated completion date is in the fall of 2018. These actions, if implemented effectively, should address the intent of our recommendation.

Federal Disaster Assistance 95

We will send copies of this chapter to the Secretary of Homeland Security, the FEMA Administrator, and the appropriate congressional committees.

Chris P. Currie
Director, Homeland Security and Justice

APPENDIX I: INDIVIDUAL ASSISTANCE DECLARATIONS REQUESTED, DECLARED, AND DENIED, CALENDAR YEARS 2008-2016

Table 9 provides the total number of Individual Assistance declaration requests made, declared, and denied, by region, state, and tribe for disaster declarations requested from calendar years 2008 through 2016.[34]

Table 9. Individual Assistance Requested, Declared, and Denied, by Region, State, and Tribe for Disaster Declarations Requested from Calendar Years 2008 through 2016[a]

Region	State	Requested	Declared	Denied
I		15	13	2
	Connecticut	4	3	1
	Massachusetts	3	3	0
	Maine	1	1	0
	New Hampshire	2	1	1
	Rhode Island	2	2	0
	Vermont	3	3	0
II		18	10	8
	New Jersey	6	3	3
	New York	7	4	3
	Puerto Rico	3	3	0
	U.S. Virgin Islands	2	0	2

---

[34] Throughout this report, and in accordance with the Stafford Act, "state" means any state of the United States, the District of Columbia, Puerto Rico, the Virgin Islands, Guam, American Samoa, and the Commonwealth of the Northern Mariana Islands. See 42 U.S.C. § 5122(4).

Table 9. (Continued)

Region	State	Requested	Declared	Denied
III		26	13	13
	Washington, D.C.	1	0	1
	Maryland	4	1	3
	Pennsylvania	3	2	1
	Virginia	6	2	4
	West Virginia	12	8	4
IV		69	53	16
	Alabama	9	6	3
	Florida	11	7	4
	Georgia	7	6	1
	Kentucky	12	10	2
	Mississippi	15	12	3
	North Carolina	5	4	1
	South Carolina	3	2	1
	Tennessee	7	6	1
V		32	16	16
	Bad River Band of Chippewa	1	0	1
	Illinois	11	6	5
	Indiana	8	5	3
	Michigan	3	1	2
	Minnesota	5	2	3
	Ohio	2	0	2
	Wisconsin	2	2	0
VI		59	32	27
	Arkansas	11	9	2
	Louisiana	14	5	9
	Navajo Nation	1	0	1
	New Mexico	1	0	1
	Oklahoma	17	9	8
	Texas	15	9	6
VII		25	13	12
	Iowa	3	3	0
	Missouri	16	8	8
	Nebraska	6	2	4
VIII		18	8	10
	Colorado	5	2	3
	Fort Peck Indian Reservation	1	0	1
	Montana	1	1	0
	North Dakota	2	2	0

Federal Disaster Assistance

Region	State	Requested	Declared	Denied
VIII	Oglala Sioux Tribe	1	1	0
	South Dakota	6	1	5
	Utah	1	0	1
	Wyoming	1	1	0
IX		18	6	12
	American Samoa	2	1	1
	Arizona	1	0	1
	California	8	2	6
	Guam	2	0	2
	Hawaii	3	1	2
	Northern Mariana Islands	1	1	0
	Nevada	1	1	0
X		14	4	10
	Alaska	5	2	3
	Colville Reservation	1	0	1
	Oregon	3	0	3
	Spokane Tribe	1	0	1
	Washington	4	2	2
Total		294	168	126

Source: GAO analysis of Federal Emergency Management Agency data. I GAO-18-366.

[a] Throughout this chapter, and in accordance with the Stafford Act, "state" means any state of the United States, the District of Columbia, Puerto Rico, the Virgin Islands, Guam, American Samoa, and the Commonwealth of the Northern Mariana Islands. See 42 U.S.C. § 5122(4).

APPENDIX II: INDIVIDUAL ASSISTANCE ACTUAL OBLIGATIONS FOR DECLARATIONS MADE, CALENDARS YEARS 2008-2016

Table 10 provides Federal Emergency Management Agency's (FEMA) Individual Assistance (IA) actual obligations for declarations made from calendar years 2008 through 2016 by state and type of disaster.[35]

---

[35] Throughout this report, actual obligations refer to funds obligated by FEMA and are not adjusted for inflation. Also, in accordance with the Stafford Act, "state" means any state of the United States, the District of Columbia, Puerto Rico, the Virgin Islands, Guam, American Samoa, and the Commonwealth of the Northern Mariana Islands. See 42 U.S.C. § 5122(4)

Table 10. Federal Emergency Management Agency's Individual Assistance (IA) Actual Obligations by State and Type of Disaster for Declarations Made from Calendar Years 2008 through 2016[a]

State	Region	Disaster type	Individuals and Households Assistance ($)	Crisis Counseling ($)	Unemployment Assistance ($)	Legal Services ($)	Disaster Case Management ($)	Total ($)
Alaska	X	Flood	11,324,269.69	37,577.00	21,263.78	3,200.00	IA program not identified[b]	11,386,310.47
Alabama	IV	Severe storm(s)	121,080,077.04	5,759,053.27	480,242.64	5,343.20	10,184,689	137,509,405.37
Arkansas	VI	Severe storm(s); Tornado	54,433,043.84	92,880.11	333,733.60	0.00	IA program not identified	54,859,657.55
		Severe storm(s)	52,041,604.92	70,191.77	324,075.94	0.00	IA program not identified	52,435,872.63
		Tornado	2,391,438.92	22,688.34	9,657.66	0.00	IA program not identified	2,423,784.92
American Samoa	IX	Earthquake	30,934,621.19	1,077,116.70	230,409.00	403.26	IA program not identified	32,242,550.15
California	IX	Fire; Earthquake	44,243,382.00	1,522,770.73	1,151,577.00	0.00	IA program not identified	46,917,729.73
		Fire	33,179,542.09	1,522,770.73	1,151,577.00	0.00	IA program not identified	35,853,889.82
		Earthquake	11,063,839.91	IA program not identified	IA program not identified	IA program not identified	IA program not identified	11,063,839.91
Colorado	VIII	Severe storm(s); Fire; Flood	68,635,065.72	7,224,031.50	711,567.62	25,753.85	2,568,072.15	79,164,490.84
		Severe storm(s)	360,885.29	457,605.11	22,640.27	0.00	IA program not identified	841,130.67

State	Region	Disaster type	Individuals and Households Assistance ($)	Crisis Counseling ($)	Unemployment Assistance ($)	Legal Services ($)	Disaster Case Management ($)	Total ($)
Colorado	VIII	Fire	No IA program identified	1,337,442.37	197,096.49	No IA program identified	No IA program identified	1,534,538.86
		Flood	68,274,180.43	5,428,984.02	491,830.86	25,753.85	2,568,072.15	76,788,821.31
Connecticut	I	Severe storm(s); Hurricane; Snow	30,397,368.44	IA program not identified	55,704.00	,314.64	423,708.48	30,878,095.56
		Severe storm(s)	5,189,469.42	IA program not identified	11,373.00	0.00	IA program not identified	5,200,842.42
		Hurricane	25,207,472.56	IA program not identified	44,331.00	1,314.64	423,708.48	25,676,826.68
		Snow	426.46	IA program not identified	IA program not identified	IA program not identified	IA program not identified	426.46
Florida	IV	Severe storm(s); Hurricane	117,901,516.08	2,596,762.51	1,051,045.81	5,000.00	406,958.67	121,961,283.07
		Severe storm(s)	88,608,589.22	2,163,480.51	523,617.81	0.00	406,958.67	91,702,646.21
		Hurricane	29,292,926.86	433,282.00	527,428.00	5,000.00	IA program not identified	30,258,636.86
Georgia	IV	Severe storm(s); Hurricane; Severe ice storm	76,554,929.43	467,614.73	375,087.29	0.00	IA program not identified	77,397,631.45
		Severe storm(s)	69,951,478.43	350,807.73	252,015.29	0.00	IA program not identified	70,554,301.45
		Hurricane	6,595,983.00	116,807.00	123,072.00	0.00	IA program not identified	6,835,862.00
		Severe ice storm	7,468.00					7,468.00

Table 10. (Continued)

State	Region	Disaster type	Individuals and Households Assistance ($)	Crisis Counseling ($)	Unemployment Assistance ($)	Legal Services ($)	Disaster Case Management ($)	Total ($)
Georgia	IV			IA program not identified	IA program not identified	IA program not identified	IA program not identified	
Hawaii	IX	Flood	3,480,427.34	595,972.17	6,510.86	0.00	IA program not identified	4,082,910.37
Iowa	VII	Severe storm(s); Flood	197,796,561.25	3,987,439.72	8,737,371.45	4,769.19	IA program not identified	210,526,141.61
		Severe storm(s)	193,012,743.41	3,467,467.04	8,640,905.13	4,769.19	IA program not identified	205,125,884.77
		Flood	4,783,817.84	519,972.68	96,466.32	0.00	IA program not identified	5,400,256.84
Illinois	V	Severe storm(s); Flood; Tornado; Snow	621,025,781.84	141,055.45	633,950.63	0.00	IA program not identified	621,800,787.92
		Severe storm(s)	455,030,045.67	23,421.00	469,600.01	0.00	IA program not identified	455,523,066.68
		Flood	163,462,430.37	IA program not identified	109,197.20	0.00	IA program not identified	163,571,627.57
		Tornado	2,530,838.68	117,634.45	55,153.42	0.00	IA program not identified	2,703,626.55
		Snow	2,467.12	IA program not identified	IA program not identified	IA program not identified	IA program not identified	2,467.12

State	Region	Disaster type	Individuals and Households Assistance ($)	Crisis Counseling ($)	Unemployment Assistance ($)	Legal Services ($)	Disaster Case Management ($)	Total ($)
Indiana	V	Severe storm(s)	102,526,308.44	1,195,111.95	319,921.63	87.20	IA program not identified	104,041,429.22
Kentucky	IV	Severe storm(s); Flood; Severe ice storm	93,515,969.88	576,468.33	581,384.81	0.00	403,453.69	95,077,276.71
		Severe storm(s)	89,711,214.12	576,468.33	490,548.81	0.00	403,453.69	91,181,684.95
		Flood	3,798,728.88	IA program not identified	90,836.00	IA program not identified	IA program not identified	3,889,564.88
		Severe ice storm	6,026.88	IA program not identified	IA program not identified	IA program not identified	IA program not identified	6,026.88
Louisiana	VI	Hurricane; Flood	1,917,500,829.58	31,530,368.43	7,448,139.15	21,420.63	46,743,951.16	2,003,244,708.95
		Hurricane	619,027,722.59	19,686,788.74	4,735,827.00	14,127.27	6,858,970.00	650,323,435.60
		Flood	1,298,473,106.99	11,843,579.69	2,712,312.15	7,293.36	39,884,981.16	1,352,921,273.35
Massachusetts	I	Severe storm(s); Tornado; Hurricane	68,363,204.51	1,452,287.62	423,365.91	335.00	212,882.76	70,452,075.80
		Severe storm(s)	58,317,381.21	676,775.41	114,375.91	0.00	IA program not identified	59,108,532.53
		Tornado	4,375,854.67	775,512.21	263,580.00	335.00	212,882.76	5,628,164.64
		Hurricane	5,669,968.63	IA program not identified	45,410.00	0.00	IA program not identified	5,715,378.63
Maryland	III	Hurricane	2,581,835.21	IA program not identified	67,794.00	0.00	IA program not identified	2,649,629.21

Table 10. (Continued)

State	Region	Disaster type	Individuals and Households Assistance ($)	Crisis Counseling ($)	Unemployment Assistance ($)	Legal Services ($)	Disaster Case Management ($)	Total ($)
Maine	I	Flood	1,220,974.02	IA program not identified	9,002.00	IA program not identified	IA program not identified	1,229,976.02
Michigan	V	Flood	148,199,521.59	IA program not identified	108,980.00	0.00	IA program not identified	148,308,501.59
Minnesota	V	Severe storm(s); Flood	5,302,057.10	0.00	230,034.00	0.00	0.00	5,532,091.10
		Severe storm(s)	2,395,457.10	IA program not identified	230,034.00	IA program not identified	IA program not identified	2,625,491.10
		Flood	2,906,600.00	IA program not identified	IA program not identified	IA program not identified	IA program not identified	2,906,600.00
Missouri	VII	Flood; Severe storm(s)	103,143,107.28	6,062,101.51	971,188.62	5,000.00	6,099,407.32	116,280,804.73
		Flood	15,988,361.87	1,348,262.56	80,939.81	5,000.00	2,603,773.00	20,026,337.24
		Severe storm(s)	87,154,745.41	4,713,838.95	890,248.81	0.00	3,495,634.32	96,254,467.49
Mississippi	IV	Severe storm(s); Hurricane; Flood	80,123,535.08	17,617.62	638,572.83	3,195.00	146,376.91	80,929,297.44
		Severe storm(s)	33,219,800.59	17,617.62	356,286.05	3,195.00	146,376.91	33,743,276.17
		Hurricane	24,432,367.39	IA program not identified	73,137.78	0.00	IA program not identified	24,505,505.17
		Flood	22,471,367.10	IA program not identified	209,149.00	0.00	IA program not identified	22,680,516.10

State	Region	Disaster type	Individuals and Households Assistance ($)	Crisis Counseling ($)	Unemployment Assistance ($)	Legal Services ($)	Disaster Case Management ($)	Total ($)
Montana	VIII	Severe storm(s)	6,610,866.31	IA program not identified	23,972.00	IA program not identified	IA program not identified	6,634,838.31
Nebraska	VII	Severe storm(s); Flood	5,777,326.34	292,823.51	90,070.50	0.00	IA program not identified	6,160,220.35
		Severe storm(s)	1,502,256.34	IA program not identified	13,561.50	IA program not identified	IA program not identified	1,515,817.84
		Flood	4,275,070.00	292,823.51	76,509.00	0.00	IA program not identified	4,644,402.51
New Hampshire	I	Hurricane	1,236,236.91	IA program not identified	24,393.00	IA program not identified	IA program not identified	1,260,629.91
New Jersey	II	Severe storm(s); Hurricane	626,445,681.49	10,808,355.17	7,493,369.94	0.00	11,586,610.46	656,334,017.06
		Severe storm(s)	16,534,228.93	IA program not identified	33,164.00	0.00	IA program not identified	16,567,392.93
		Hurricane	609,911,452.56	10,808,355.17	7,460,205.94	0.00	11,586,610.46	639,766,624.13
Nevada	IX	Severe storm(s)	1,893,834.06	IA program not identified	IA program not identified	IA program not identified	IA program not identified	1,893,834.06
New York	II	Hurricane; Severe storm(s)	1,190,712,765.04	54,528,126.89	17,719,931.04	269.68	46,907,949.83	1,309,869,042.48
		Hurricane	1,127,960,708.68	51,892,473.89	16,721,106.79	269.68	45,692,130.83	1,242,266,689.87
		Severe storm(s)	62,752,056.36	2,635,653.00	998,824.25	0.00	1,215,819.00	67,602,352.61

Table 10. (Continued)

State	Region	Disaster type	Individuals and Households Assistance ($)	Crisis Counseling ($)	Unemployment Assistance ($)	Legal Services ($)	Disaster Case Management ($)	Total ($)
North Carolina	IV	Severe storm(s); Hurricane; Mud/ Land Slide	165,714,736.18	1,683,795.10	2,523,839.68	5,000.00	4,521,340.58	174,448,711.54
		Severe storm(s)	14,125,060.36	986,839.61	183,754.82	0.00	IA program not identified	15,295,654.79
		Hurricane	151,589,675.82	696,955.49	2,340,084.86	5,000.00	4,521,340.58	159,153,056.75
		Mud/ Landslide	0.00	IA program not identified	IA program not identified	IA program not identified	IA program not identified	0.00
North Dakota	VIII	Severe storm(s); Flood	194,977,101.05	2,009,270.33	3,898,889.69	5,000.00	IA program not identified	200,890,261.07
		Severe storm(s)	8,216,818.62	IA program not identified	1,337,457.60	0.00	IA program not identified	9,554,276.22
		Flood	186,760,282.43	2,009,270.33	2,561,432.09	5,000.00	IA program not identified	191,335,984.85
Northern Mariana Islands	IX	Typhoon	30,764,794.87	1,136,694.74	697,013.00	5,000.00	1,677,942.40	34,281,445.01
Oklahoma	VI	Severe storm(s); Fire; Tornado	51,131,095.99	2,547,961.19	388,873.33	5,000.00	3,534,616.00	57,607,546.51
		Severe storm(s)	27,896,658.64	1,451,811.43	191,232.33	IA program not identified	3,534,616.00	33,074,318.40
		Fire	8,945,183.13	IA program not identified	23,000.00	0.00	IA program not identified	8,968,183.13

State	Region	Disaster type	Individuals and Households Assistance ($)	Crisis Counseling ($)	Unemployment Assistance ($)	Legal Services ($)	Disaster Case Management ($)	Total ($)
Oklahoma	VI	Tornado	14,289,254.22	1,096,149.76	174,641.00	5,000.00	IA program not identified	15,565,044.98
Pennsylvania	III	Hurricane; Flood	162,245,344.66	801,143.93	1,895,308.00	0.00	544,836.71	165,486,633.30
		Hurricane	41,084,124.81	IA program not identified	147,840.00	0.00	IA program not identified	41,231,964.81
		Flood	121,161,219.85	801,143.93	1,747,468.00	0.00	544,836.71	124,254,668.49
Puerto Rico	II	Hurricane; Severe storm(s)	52,532,483.18	5,659,880.22	45,202.15	IA program not identified	IA program not identified	58,237,565.55
		Hurricane	30,299,140.51	2,079,175.92	39,309.00	IA program not identified	IA program not identified	32,417,625.43
		Severe storm(s)	22,233,342.67	3,580,704.30	5,893.15	IA program not identified	IA program not identified	25,819,940.12
Rhode Island	I	Hurricane; Severe storm(s)	36,689,593.97	1,786,157.00	355,464.54	0.00	IA program not identified	38,831,215.51
		Hurricane	410,938.22	78,724.00	17,294.88	IA program not identified	IA program not identified	506,957.10
		Severe storm(s)	36,278,655.75	1,707,433.00	338,169.66	0.00	IA program not identified	38,324,258.41
South Carolina	IV	Hurricane; Flood	128,587,779.98	6,802,962.20	1,107,267.00	5,000.00	11,442,606.00	147,945,615.18
		Hurricane	39,663,704.61	2,175,297.41	173,462.00	5,000.00	4,553,144.00	46,570,608.02
		Flood	88,924,075.37	4,627,664.79	933,805.00	0.00	6,889,462.00	101,375,007.16

Table 10. (Continued)

State	Region	Disaster type	Individuals and Households Assistance ($)	Crisis Counseling ($)	Unemployment Assistance ($)	Legal Services ($)	Disaster Case Management ($)	Total ($)
South Dakota	VIII	Flood; Severe storm(s)	55,416,328.60	IA program not identified	87,034.10	0.00	IA program not identified	55,503,362.70
		Flood	4,753,534.43	IA program not identified	69,252.45	0.00	IA program not identified	4,822,786.88
		Severe storm(s)	50,662,794.17	IA program not identified	17,781.65	IA program not identified	IA program not identified	50,680,575.82
Tennessee	IV	Severe storm(s); Fire	194,654,226.96	3,919,165.75	866,622.14	7,935.66	IA program not identified	*199,447,950.51*
		Severe storm(s)	190,690,926.96	3,797,041.75	866,622.14	2,935.66	IA program not identified	195,357,526.51
		Fire	3,963,300.00	122,124.00	IA program not identified	5,000.00	IA program not identified	4,090,424.00
Texas	VI	Hurricane; Flood; Severe storm(s); Fire	1,104,762,457.87	17,201,156.42	10,175,299.36	10,000.00	17,480,757.00	*1,149,629,670.65*
		Hurricane	887,157,346.07	7,300,332.39	8,384,537.36	0.00	IA program not identified	902,842,215.82
		Flood	128,976,403.78	3,611,909.01	586,328.00	10,000.00	10,271,993.00	143,456,633.79
		Severe storm(s)	73,113,228.76	5,803,009.18	840,459.00	0.00	7,208,764.00	86,965,460.94
		Fire	15,515,479.26	485,905.84	363,975.00	0.00	IA program not identified	16,365,360.10
Virginia	III	Hurricane; Earthquake	26,453,711.38	68,664.15	8,210.61	0.00	IA program not identified	*26,530,586.14*

State	Region	Disaster type	Individuals and Households Assistance ($)	Crisis Counseling ($)	Unemployment Assistance ($)	Legal Services ($)	Disaster Case Management ($)	Total ($)
		Hurricane	6,026.88	IA program not identified	IA program not identified	IA program not identified	IA program not identified	6,026.88
		Earthquake	16,566,971.42	68,664.15	1,346.61	0.00	IA program not identified	16,636,982.18
Virgin Islands	II	Hurricane; Severe storm(s)	2,136.07	IA program not identified	IA program not identified	IA program not identified	IA program not identified	2,136.07
		Hurricane	656.07	IA program not identified	IA program not identified	IA program not identified	IA program not identified	656.07
		Severe storm(s)	1,480.00	IA program not identified	IA program not identified	IA program not identified	IA program not identified	1,480.00
Vermont	I	Hurricane; Severe storm(s)	26,557,306.49	1,087,593.46	494,750.04	0.00	2,222,641.00	30,362,290.99
		Hurricane	23,398,923.48	1,087,593.46	469,905.03	0.00	2,222,641.00	27,179,062.97
		Severe storm(s)	3,158,383.01	IA program not identified	24,845.01	0.00	IA program not identified	3,183,228.02
Washington	X	Flood; Mud/ Landslide	10,964,235.08	269,573.00	2,215,653.00	848.15	IA program not identified	13,450,309.23
		Flood	9,240,849.30	IA program not identified	1,620,153.00	848.15	IA program not identified	10,861,850.45
		Mud/ Landslide	1,723,385.78	269,573.00	595,500.00	0.00	IA program not identified	2,588,458.78

Table 10. (Continued)

State	Region	Disaster type	Individuals and Households Assistance ($)	Crisis Counseling ($)	Unemployment Assistance ($)	Legal Services ($)	Disaster Case Management ($)	Total ($)
Wisconsin	V	Severe storm(s)	113,491,654.91	2,015,907.43	342,123.00	0.00	IA program not identified	115,849,685.34
West Virginia	III	Hurricane; Flood; Severe storm(s)	91,090,572.42	3,117,356.85	292,845.73	0.00	5,821,472.00	100,322,247.00
		Hurricane	52,093.97	IA program not identified	IA program not identified	IA program not identified	IA program not identified	52,093.97
		Flood	52,939,896.68	2,660,395.92	219,479.00	0.00	5,821,472.00	61,641,243.60
		Severe storm(s)	38,098,581.77	456,960.93	73,366.73	0.00	IA program not identified	38,628,909.43
Wyoming	VIII	Flood; Severe storm(s)	2,737,976.80	222,519.00	68,816.00	0.00	IA program not identified	3,029,311.80
		Flood	2,736,003.04	222,519.00	68,816.00	0.00	IA program not identified	3,027,338.04
		Severe storm(s)	1,973.76	IA program not identified	IA program not identified	IA program not identified	IA program not identified	1,973.76
Grand total								8,610,481,911.13

Source: GAO analysis of FEMA data. I GAO-18-366.

[a] Table includes actual obligations made from the time of the declaration to March 2017. Throughout this chapter, and in accordance with the Stafford Act, "state" means any state of the United States, the District of Columbia, Puerto Rico, the Virgin Islands, Guam, American Samoa, and the Commonwealth of the Northern Mariana Islands. See 42 U.S.C. § 5122(4).

[b] "IA program not identified" refers to those programs that were not identified in the Integrated Financial Management Information System obligations costs data.

Federal Disaster Assistance 109

APPENDIX III: REGIONAL ADMINISTRATOR'S VALIDATION AND RECOMMENDATION TEMPLATE

As part of the Federal Emergency Management Agency's (FEMA) declaration process, FEMA's regional offices are to evaluate states' or tribes' declaration requests, including the IA declaration request, and make a recommendation called the Regional Administrator's Validation and Recommendation (RVAR) and submit the RVAR to FEMA headquarters. In June 2012, FEMA headquarters issued a template for FEMA regional offices to use in developing the RVAR as identified in figure 3.[104]

REGIONAL ADMINISTRATOR'S VALIDATION AND RECOMMENDATION
MAJOR DISASTER/EMERGENCY DECLARATION REQUEST
DATE

MEMORANDUM FOR:
                 Acting Associate Administrator
                 Office of Response and Recovery

FROM:
                 Regional Administrator's Name
                 Region //

SUBJECT:
                 Regional Administrator's Validation and Recommendation
                 State of //

Please find my validation of the key elements contained in Governor ////'s form and cover letter dated /// //, ///, and my recommendation for *approving/denying* the request.

The Governor's form, including attachments satisfies legal requirements for emergency and major disaster declaration requests under 42 U.S.C. 5170 and 5191, respectively, as implemented at 44 C.F.R 206.35 and 206.36.

Provide a brief description of the event that led to the need for a joint federal, state, territory, tribal, and local government Preliminary Damage Assessment (PDA). Any additional information regarding the area or incident, such as terrain or other unusual items that would give a better understanding of the situation should be included.

The Congressional Representation for the requested counties is as follows:

Senators /// last name (/-//) party affiliation and two letter state abbreviation and /// (/-//) and Representatives /// (/-//) and /// (/-//) represent the affected areas.

VALIDATION KEY ELEMENTS
- Summary of event   Refer to Section / of Request Form & Cover Letter
- Execution of State EM Plan   Refer to Section/ of Request Form & Cover Letter
- State of Emergency   Refer to Section / of Request Form
- HM Plan   Refer to Section / of Request Form
- Damage to critical facilities   Refer to Section / of Request Form & Cover Letter
- Health and safety concerns   Refer to Section / of Request Form & Cover Letter
- Joint PDAs requested and dates   Refer to Section / of Request Form
- Severity and magnitude   Refer to Section / of Request Form
- Areas Requested   Refer to Section / of Request Form
- State and local resources   Refer to Section / of Request Form & Cover Letter
- Preliminary estimates   Refer to Enclosures of Cover Letter
- OFA estimated requirements   Refer to Enclosure C of Cover Letter
- Certification to cost share   Refer to Section / of Request Form
- DFA and debris assurances   Refer to Section / of Request Form

---

[104] Throughout this report, and in accordance with the Stafford Act, "state" means any state of the United States, the District of Columbia, Puerto Rico, the Virgin Islands, Guam, American Samoa, and the Commonwealth of the Northern Mariana Islands. See 42 U.S.C. § 5122(4).

110   *United States Government Accountability Office*

- SCO designation   Refer to Section / of Request Form
- Legal sufficiency review   Name of Regional Counsel

CHART: Complete only the applicable sections and remove shading before finalizing.

Green – All requests
Yellow – IA requests
Blue – PA requests

Type of Disaster	*Specify incident type as it should appear in the declaration letter*
Programs Requested	IA, PA, & HM
PDA Period	IA Dates:
	PA Dates:
Total Estimated Federal Obligation	$////
Recent Disasters in Same Area	*Provide the State's disaster history within the last 12 months.*
Trauma	*Indicate deaths, injuries, power outages, and disruption of other community functions and services.*
Damage Concentration	Low/Medium/High/Extreme *Provide information on the concentration of damages to individuals and households.*
Homes Destroyed	///
Homes Major Damage	///
Homes Minor Damage	///
Homes Affected	////
Ownership	//./%
Insurance	//./% *Provide percentage of applicable insurance coverage depending on the type*

Figure 3. (Continued)

Federal Disaster Assistance

	of damage such as such as, homeowners, flood, earthquake, etc...
Flood Insurance (if applicable)	*//.J% *For flood disasters, include the percentage of National Flood Insurance Program (NFIP) coverage in the affected areas.*
Low Income	*//.J%*
Median Household Income	*$/////*
Poverty	*//.J%*
Disabled	*//.J%*
Elderly	*//.J%*
SBA Assistance	*Provide the estimates obtained from SBA for the number of residential and business loans expected and the estimated total cost.* SBA officials reported that under the Agency's disaster loan programs, (number) residential loans might be available for a total program cost of $(amount) and (number) business loans estimated at a total cost of $(amount).
Level of Volunteer Assistance	None/Scarce/Moderate/Robust
Volunteer Agency Assistance	*Describe voluntary agency assistance provided to the community in anticipation of or as a result of the event. (including but not limited to activities taken by the American Red Cross (ARC), Salvation Army, Southern Baptists, and Voluntary Organizations Active in Disaster; voluntary agency and/or municipal shelter information, including the number of shelters open, the peak population, total number of overnight stays, and when the shelters closed; number of mental health contacts made by voluntary agencies; number of ARC cases open and closed; number of clean up kits provided; number*

Figure 3. (Continued)

112 *United States Government Accountability Office*

	of meals served; number of fixed and/or mobile feeding sites; level of ARC operation (I-V); and any other type of assistance that was provided by voluntary organizations). The American Red Cross opened // shelters housing /// individuals. All individuals are back in their homes, staying with friends, or have been placed in hotels. All shelters are now closed.
Statewide Per Capita	$////
Countywide Per Capita/s	////// County ($////), etc....
Localized Impacts	*Indicate if any local communities have sustained significant damage and have incurred extremely high per capita impacts, provide that information, especially if the localized damage is in the tens or even hundreds of dollars per capita.*
Insurance Coverage in Force	--describe--- Infrastructure damage known to be insured is not included in the total eligible Public Assistance cost estimate.
Hazard Mitigation Measures	*Indicate date of State Mitigation Plan (SMP) approval, and type of SMP (Standard or Enhanced), and corresponding percentage for HMGP calculation (15%) OR (20%).* *Describe State and local government measures that contributed to the reduction of disaster damages for the disaster under consideration.*
Other Federal Assistance Available	*Provide information about damage to public facilities eligible under authorities other than the Stafford Act, such as, Federal-aid-system roads (Federal Highway Administration [FHWA]), water control facilities (U.S. Army Corps of*

Figure 3. (Continued)

	Engineers [USACE] or Natural Resource Conservation Service [NRCS]).

INDIVIDUAL ASSISTANCE

Provide a brief description of the impact of the event on individuals and households in the affected areas, including evaluating trauma, concentration of damage, impact on populations with greater need, and other available assistance. Provide information from the Individual Assistance Preliminary Damage Assessment (PDA), including method of assessment, areas surveyed, number of homes affected, degree of residential damage, ownership, insurance, low-income, and any other significant needs. For flood disasters, provide the percentage of homes affected that are in the Special Flood Hazard Area (SFHA), the number of homes that are subject to a obtain and maintain flood insurance requirement, and number of flood insurance claims filed if available. Discuss unmet needs and requirements for assistance. Provide justification to support recommendation to approve or deny the Individual Assistance program(s). Provide clear explanation of the methodology for determining the extent of damage and level of insurance coverage when recommending Individual Assistance for expedited requests.

PUBLIC ASSISTANCE

Provide a brief description of the impact of the event on public facilities and eligible private nonprofits, as well as the costs associated with eligible debris removal and emergency protective measures. Provide a brief analysis of infrastructure damage by category, including the cost estimates for each category and the percentage of the total eligible Public Assistance cost estimate. Include in your discussion significant impacts to specific areas. Provide justification to support recommendation to approve or deny the Public Assistance program.

Category A (Debris Removal) – *IIIII.*

Category B (Protective Measures) – *IIIII.*

Category C (Roads and Bridges) – *IIIII.*

Category D (Water Control Facilities) – *IIIII.*

Category E (Buildings and Equipment) – *IIIII.*

Category F (Utilities) – *IIIII. (Please include confirmation that there is no private utility involvement and if there is please note and separate the private and public utilities impacted.)*

Category G (Other (recreation, etc.)) – *IIIII. (Please confirmation that these are municipal/publicly owned.)*

Figure 3. (Continued)

114 *United States Government Accountability Office*

<u>RECOMMENDATION</u>

(Please include any pre-decisional information or comments in this section only)

- I recommend the Governor's request be granted.

- I recommend an incident period of ///.

  The incident type and period should be stated as it will appear in the declaration. It may vary somewhat from the information in the Governor's form and letter. If there is a great disparity, state why and provide NWS Summary. If the event is continuing at the time this report is being prepared, the closing date may be left open and closed later.

- In the event of a declaration, I recommend that *Individual Assistance* and *Public Assistance* be made available in the following jurisdictions:

  > The counties of ////, ////, ////, ////, ///, ///, ///, ////, ///, and /// for Individual Assistance. /// County is not recommended for Individual Assistance.

  > The counties of ////, ////, ////, ////, ///, ///, and //// for Public Assistance.

  > The counties of ///, ///, ///, and //// are not recommended for Public Assistance.

  > I recommend direct federal assistance.

- In the event of a declaration, the Hazard Mitigation Grant Program would be available *statewide OR list specific areas*, in accordance with the State's request and subject to Local Mitigation Plan requirements identified at 44 C.F.R. § 206.434(b).

- In the event of a declaration, I recommend //// be designated as the Federal Coordinating Officer.

<u>ATTACHMENTS</u>

State Map: Please send as a separate pdf attachment with a map, indicating the requested areas and programs. *See pdf map example below, containing a legend of the requested counties and programs

PDA Spreadsheet: Please note that the PA dollar amounts and Category totals may vary somewhat from the dollar amounts in the Governor's spreadsheet. If there is any disparity in the dollar amounts, please explain the disparity in the narrative portion of the RVAR. Some examples may include ineligible, insurable, or unverifiable costs that were backed out.

Figure 3. Federal Emergency Management Agency's (FEMA) Regional Administrator's Validation and Recommendation Template.

Federal Disaster Assistance 115

APPENDIX IV: INFORMATION ON THE ELEMENTS OF THE SIX INDIVIDUAL ASSISTANCE REGULATORY FACTORS

Tables 11 through 16 provide information on each element of the 6 Individual Assistance (IA) regulatory factors documented in the Regional Administrator's Validation and Recommendation (RVAR) from July 2012 through December 2016 by the Federal Emergency Management Agency region.

Table 11. Information on Concentration of Damages by Element Found in Regional Administrator's Validation and Recommendation (RVAR) from July 2012 through December 2016

Factor: Concentration of Damages		2012 RVARs	2013 RVARs	2014 RVARs	2015 RVARs	2016 RVARs
Damage to critical facilities	Documented	6	13	3	11	12
	Not documented	1	0	11	11	13
Damage concentration	Documented	7	12	14	22	25
	Not documented	0	1	0	0	0
Homes destroyed	Documented	7	12	14	22	24
	Not documented	0	1	0	0	1
Homes with major damage	Documented	7	12	14	22	24
	Not documented	0	1	0	0	1
Homes with minor damage	Documented	7	12	14	22	24
	Not documented	0	1	0	0	1
Homes affected	Documented	7	12	14	22	24
	Not documented	0	1	0	0	1
Total number of RVARs by year		7	13	14	22	25

Source: GAO analysis of Federal Emergency Management Agency's RVAR. I GAO-18-366.

116   *United States Government Accountability Office*

Table 12. Information on Trauma by Element Found in Regional Administrator's Validation and Recommendation (RVAR) from July 2012 through December 2016

Factor: Trauma		2012 RVARs	2013 RVARs	2014 RVARs	2015 RVARs	2016 RVARs
Death	Documented	7	10	13	20	23
	Not documented	0	3	1	2	2
Injuries	Documented	4	7	11	17	10
	Not documented	3	6	3	5	15
Power outage	Documented	6	9	13	13	16
	Not documented	1	4	1	9	9
Disruption of community functions/services	Documented	5	12	11	14	20
	Not documented	2	1	3	8	5
Total number of RVARs by year		7	13	14	22	25

Source: GAO analysis of Federal Emergency Management Agency's RVAR. I GAO-18-366.

Table 13. Information on Special Populations by Element Found in Regional Administrator's Validation and Recommendation (RVAR) from July 2012 through December 2016

Factor: Special Populations		2012 RVARs	2013 RVARs	2014 RVARs	2015 RVARS	2016 RVARs
Low income	Documented	7	10	14	21	22
	Not documented	0	3	0	1	3
Poverty	Documented	7	12	14	22	25
	Not documented	0	1	0	0	0
Disabled	Documented	6	10	14	22	25
	Not documented	1	3	0	0	0
Elderly	Documented	6	12	14	22	25
	Not documented	1	1	0	0	0
Total number of RVARs by year		7	13	14	22	25

Source: GAO analysis of Federal Emergency Management Agency's RVAR. I GAO-18-366.

Federal Disaster Assistance

Table 14. Information on Voluntary Assistance by Element Found in Regional Administrator's Validation and Recommendation (RVAR) from July 2012 through December 2016

Factor: Voluntary Assistance		2012 RVARs	2013 RVARs	2014 RVARs	2015 RVARs	2016 RVARs
Hazard mitigation	Documented	5	13	13	21	25
	Not documented	2	0	1	1	0
Level of volunteer assistance	Documented	7	13	14	21	25
	Not documented	0	0	0	1	0
Total number of RVARs by year		7	13	14	22	25

Source: GAO analysis of Federal Emergency Management Agency's RVAR. I GAO-18-366.

Table 15. Information on Insurance Coverage Found in Regional Administrator's Validation and Recommendation (RVAR) from July 2012 through December 2016

Factor: Insurance Coverage		2012 RVARs	2013 RVARs	2014 RVARs	2015 RVARs	2016 RVARs
Home ownership	Documented	0	3	0	1	3
	Not documented	7	10	14	21	22
Insurance	Documented	6	10	13	18	24
	Not documented	1	3	1	4	1
Flood insurance (if applicable)	Documented	6	7	11	11	15
	Not documented	1	1	3	9	6
	Not applicable[a]	0	5	0	2	4
Total number of RVARs by year		7	13	14	22	25

Source: GAO analysis of Federal Emergency Management Agency's RVAR. I GAO-18-366.

118   United States Government Accountability Office

Table 16. Information on Average Amount of Individual Assistance by State Found in Regional Administrator's Validation and Recommendations (RVAR) from July 2012 through December 2016

Factor: Average Amount of Individual Assistance by State		2012 RVARs	2013 RVARs	2014 RVARs	2015 RVARS	2016 RVARs
Recent disasters in the same area in past 12 months	Documented	7	13	10	21	25
	Not documented	0	0	4	1	0
Impact and frequency of prior disasters	Documented	2	2	1	5	2
	Not documented	5	11	13	17	23
Total number of RVARs by year		7	13	14	22	25

Source: GAO analysis of Federal Emergency Management Agency's RVAR. I GAO-18-366.

APPENDIX V: COMMENTS FROM THE DEPARTMENT OF HOMELAND SECURITY

U.S. Department of Homeland Security
Washington, DC 20528

May 18, 2018

Chris P. Currie
Director, Homeland Security and Justice
U.S. Government Accountability Office
441 G Street, NW
Washington, DC 20548

Re: Management's Response to Draft Report GAO-18-366, "FEDERAL DISASTER ASSISTANCE: Individual Assistance Requests Often Granted, but FEMA Could Better Document Factors Considered"

Dear Mr. Currie:

Thank you for the opportunity to review and comment on this draft report. The U.S. Department of Homeland Security (DHS) appreciates the U.S. Government Accountability Office's (GAO) work in planning and conducting its review and issuing this report.

The Department is pleased to note GAO's positive recognition related to the Federal Emergency Management Agency's (FEMA) development of the Regional Administrator's Validation and Recommendation (RVAR) to ensure FEMA regions consistently obtain and document the information needed to make disaster declaration recommendations to the President based on Individual Assistance (IA) factors. FEMA is committed to enhancing the overall quality and completeness of documentation relating to the IA declaration process, as appropriate.

Federal Disaster Assistance

119

It is also important to note that while FEMA is aware its regions do not always address each element in the RVAR template, generally, when this occurs, there are valid reasons. For example, a particular element may not be addressed within an RVAR due to a lack of sufficient documentation or information available within the time constraints pertaining to an event. In addition, certain elements may not be addressed within an RVAR if other elements addressed within the RVAR provide sufficient information for a particular declaration request to be duly considered. Every disaster is different, and the factors considered will vary based on considerations such as size and scale of the event, state and local capability, and the specific assistance programs requested. As such, not every

element in the RVAR template may be necessary for the consideration of a declaration request.

The draft report contained one recommendation, with which the Department concurs. Attached find our detailed response to the recommendation. Technical comments were previously provided under a separate cover.

Again, thank you for the opportunity to review and comment on this draft report. Please feel free to contact me if you have any questions. We look forward to working with you in the future.

Sincerely,

JIM H. CRUMPACKER, CIA, CFE
Director
Departmental GAO-OIG Liaison Office

Attachment: Management Response to Recommendations Contained in GAO-18-366

GAO recommended that the Administrator of FEMA:

Recommendation 1: Evaluate why regions are not completing the RVARs for each element of the current IA regulatory factors and take corrective steps, if necessary.

Response: Concur. A working group of FEMA headquarters stakeholders (including a representative from the Office of Response and Recovery's [ORR] Declarations Section) will draft survey questions for response by FEMA region officials to identify the common reasons why an element of an IA regulatory factor may not be addressed within an RVAR. The working group will analyze the survey responses and assess whether additional action is necessary to address why FEMA regions are not always addressing all elements in the RVAR template. The working group will then present its findings to FEMA senior leadership and if leadership concludes that additional action is necessary, ORR will prepare and send a memorandum to the regions with additional guidance regarding the appropriate preparation of RVARs. Estimated Completion Date: October 15, 2018.

In: Issues in Disaster Recovery and Assistance ISBN: 978-1-53616-308-7
Editor: Donatien Moïse   © 2019 Nova Science Publishers, Inc.

Chapter 4

FEMA AND SBA DISASTER ASSISTANCE FOR INDIVIDUALS AND HOUSEHOLDS: APPLICATION PROCESS, DETERMINATIONS, AND APPEALS[*]

Bruce R. Lindsay and Shawn Reese

SUMMARY

The Federal Emergency Management Agency's (FEMA's) Individual Assistance (IA) program and the Small Business Administration's (SBA's) Disaster Loan Program are the federal government's two primary sources of financial assistance to help individuals and households recover and rebuild from a major disaster. In many cases, disaster survivors find that they need assistance from both of these programs in addition to other sources of assistance including private insurance, state and local government assistance, and assistance from private voluntary organizations.

---

[*] This is an edited, reformatted and augmented version of Congressional Research Service, Publication No. R45238, dated June 22, 2018.

Though FEMA IA and the SBA Disaster Loan Program are separate programs administered by different agencies, in many ways they are interconnected. SBA and FEMA share real-time data on disaster grant and loan approvals to identify potential duplication of benefits while providing individuals and households with federal assistance that can be used in conjunction with each other to meet recovery needs. The two programs are also interconnected in the way they are administered to determine loan and grant eligibility. Furthermore, eligibility and assistance from one source can affect eligibility and assistance from the other source.

It could be argued the overlap between the two programs provides an effective means to identify duplication and provide federal assistance; however, the overlap also causes some confusion. Some in Congress are concerned that elements of the application process are not entirely known. For instance, it is unclear to some what criteria are used to determine assistance eligibility as well as how decisions are made with respect to whether an applicant should be provided a grant or a loan (or both). It is also unclear whether FEMA and SBA determine eligibility on a case-by-case basis, or if eligibility criteria are applied uniformly.

This chapter provides an overview of the two programs including discussions about

- how declarations put the programs into effect;
- the application process for both programs;
- the criteria used by FEMA and the SBA to determine assistance; and
- the FEMA and SBA appeal processes.

The chapter concludes with policy observations and considerations for Congress.

INTRODUCTION

Individuals and households that suffer uninsured or underinsured losses under a major disaster declaration typically apply for Individual Assistance (IA), administered by the Federal Emergency Management Agency (FEMA), and may also apply for disaster loans, administered by

FEMA and SBA Disaster Assistance for Individuals ... 123

the Small Business Administration (SBA).[1] This chapter opens with an overview of the two programs and a discussion about how declarations are used to put them into effect. The report also discusses their respective application processes and eligibility criteria used by FEMA and SBA to make grant and loan determinations, respectively. The report then describes the appeals process before concluding with policy observations and considerations.

FEMA IA and the SBA Disaster Loan Program are interlaced to a certain degree. Functionally, SBA and FEMA have a computer matching agreement (CMA) to share real-time data on assistance provided to applicants. SBA and FEMA use the interface between their systems to identify potential duplication of benefits (DOB) and determine loan and grant eligibility. From an administrative perspective, eligibility and assistance from one source can impact eligibility and assistance from the other source. While the overlap between the two programs may have some benefits, it arguably also causes some confusion. Moreover, as some observers have pointed out, there are elements of the application process that are not entirely known or understood. For instance, Members have asked for clarification concerning the criteria used to determine eligibility for grants and/or loans, as well as how decisions are made with respect to whether an applicant is provided a grant, loan, or both. Others have questioned whether determinations are made on a case-by-case basis, or if the determination criteria are applied uniformly to all applicants seeking disaster assistance.

OVERVIEW OF PROGRAMS

The following sections provide descriptions of the SBA and FEMA programs that provide assistance to individuals and households. In many

---

[1] For more information about IA, see CRS Report R45085, *FEMA Individual Assistance Programs: In Brief*, by Shawn Reese. For more information on SBA disaster loans, see CRS Report R41309, *The SBA Disaster Loan Program: Overview and Possible Issues for Congress*, by Bruce R. Lindsay; and CRS Report R44412, *SBA Disaster Loan Program: Frequently Asked Questions*, by Bruce R. Lindsay.

cases, disaster survivors find that they need assistance from both of these programs in addition to other sources of assistance including private insurance, state and local government assistance, and assistance from private voluntary organizations in order to fully recover.

SBA Home Disaster Loans

In addition to small business disaster loans, homeowners, renters, and personal property owners located in a declared disaster area are also eligible to apply for an SBA home disaster loan. SBA home disaster loans can be conceptualized as two categories of loans according to how the proceeds are put to use: Personal Property Loans and Real Property Loans.

Personal Property Loans

A Personal Property Loan provides a creditworthy homeowner or renter located in a declared disaster area with up to $40,000 to repair or replace personal property owned by the victim.[2]

Eligible items include furniture, appliances, clothing, and automobiles damaged or lost in a disaster. These loans cover only uninsured or underinsured property and primary residences in a declared disaster area. Eligibility of luxury items with functional use, such as antiques and rare artwork, is limited to the cost of an ordinary item meeting the same functional purpose. Interest rates for Personal Property Loans cannot exceed 8% per annum, or 4% per annum if the applicant is found by SBA to be unable to obtain credit elsewhere. Generally, borrowers pay equal monthly installments of principal and interest, beginning five months from the date of the loan. Loan maturities may be up to 30 years.

Real Property Loans

Real Property Loans provide creditworthy homeowners with uninsured or underinsured loss located in a declared disaster area with up to $200,000

---

[2] 13 C.F.R. §123.105(a)(1).

FEMA and SBA Disaster Assistance for Individuals ...   125

to repair or replace the homeowner's primary residence to its predisaster condition.[3] The loans may not be used to upgrade a home or build additions to the home, unless the upgrade or addition is required by city or county building codes such as a code-required elevation. Repair or replacement of landscaping and/or recreational facilities cannot exceed $5,000. A homeowner may borrow funds to cover the cost of improvements to protect their property against future damage (e.g., elevation, retaining walls, sump pumps, etc.). Mitigation funds may not exceed 20% of the disaster damage, as verified by SBA, to a maximum of $200,000 for home loans.[4] As previously mentioned, interest rates cannot exceed 8% per annum, or 4% per annum if the applicant is unable to obtain credit elsewhere. Generally, borrowers pay equal monthly installments of principal and interest, beginning five months from the date of the loan. Loan maturities may be up to 30 years.

FEMA Individual Assistance

IA can include several programs, depending on whether the governor of the affected state or the tribal leader requests that specific type of FEMA assistance. FEMA's IA includes (1) Mass Care and Emergency Assistance, (2) the Crisis Counseling Assistance and Training Program, (3) Disaster Unemployment Assistance, (4) Disaster Legal Services, (5) Disaster Case Management, and (6) the Individuals and Households Program.[5]

Mass Care and Emergency Assistance

Mass Care includes directly supporting sheltering, feeding, emergency supply distribution, and family reunification. Emergency Assistance includes a variety of services and functions, including coordination of

---

[3] 13 C.F.R. §123.105(a)(2).

[4] 13 C.F.R. §123.107.

[5] For more information on FEMA IA programs see CRS Report R45085, *FEMA Individual Assistance Programs: In Brief*, by Shawn Reese. See also FEMA, "Individual Assistance Program Tools," https://www.fema.gov/individual-assistance-program-tools. P.L. 93-288, Title IV, §§ 401- 426.

126

Bruce R. Lindsay and Shawn Reese

volunteer organizations and unsolicited donations, managing unaffiliated volunteers and community relief services, supporting transitional sheltering, and supporting mass evacuations.

Crisis Counseling Assistance and Training Program

The Crisis Counseling Assistance and Training Program assists individuals and communities recovering from the effects of a disaster through community-based outreach and psycho-educational services.[6] The program supports short-term counseling of disaster survivors. The program also provides information on coping strategies and emotional support by linking the survivor with other individuals and agencies that help survivors in the recovery process.

Disaster Unemployment Assistance

Disaster Unemployment Assistance provides information and resources to individuals who were employed or self-employed, or were scheduled to begin employment during a disaster. It may also be provided to those who can no longer work or perform their job duties due to damage to their place of employment, who do not qualify for regular unemployment benefits from a state, or who cannot perform work or self-employment due to an injury as direct result of a disaster.[7]

Disaster Legal Services

Disaster Legal Services provides legal assistance to low-income individuals who are unable to secure adequate legal services that meet their disaster-related needs.

Disaster Case Management

Disaster Case Management provides a partnership between a case manager and the disaster survivor to assist them in carrying out a disaster

---

[6] Psycho-educational service consists of therapeutic treatment for disaster victims that provides information and support to help them better understand and cope with their situation.

[7] For more information on disaster unemployment assistance, see CRS Report RS22022, *Disaster Unemployment Assistance (DUA)*, by Julie M. Whittaker.

recovery plan. The recovery plan includes resources, service, decisionmaking priorities, progress reports, and the goals needed to close their case.

Individuals and Households Program (IHP)

The Individuals and Households Program (IHP) is comprised of two categories of assistance: Housing Assistance, and Other Needs Assistance. Housing Assistance may include financial assistance to

- reimburse for hotels, motels, or other short-term lodging.
- rent alternate housing accommodations while the applicant is displaced from their primary residence.
- make repairs to primary residence.
- assist in replacing owner-occupied residences when the residence is destroyed.
- enter into lease agreements with owners of multifamily rental properties located in the disaster area.

Housing Assistance may also include home repair and construction services provided in insular areas outside the continental United States and other locations where no alternative housing resources are available and where types of FEMA housing assistance that are normally provided (such as rental assistance) are unavailable, infeasible, or not cost-effective. In addition, FEMA may provide manufactured housing units as a form of temporary housing through its Transitional Sheltering Assistance program.

The Robert T. Stafford Disaster Relief and Emergency Assistance Act (P.L. 93-288, as amended, hereinafter the Stafford Act) originally capped IA financial assistance at "no greater than \$25,000 under this section with respect to a single major disaster."[8] The Stafford Act was later amended to adjust the cap annually to reflect changes in the Consumer Price Index published by the Department of Labor.[9] The current IA cap is \$33,300 and

---

[8] Stafford Act, §408(h)(1).
[9] Stafford Act, §408(h)(2).

the approximate per household or individual award amount is \$8,500.[10] Households that have damages exceeding \$33,300 may also need an SBA disaster loan to rebuild and repair their home.

Other Needs Assistance (ONA) provides financial assistance for other disaster-related expenses and needs. ONA is divided into two categories: (1) SBA dependent, and (2) non-SBA dependent.

SBA Dependent ONA

Only individuals or households who do not qualify for a loan from the SBA may be eligible for the following types of assistance:

- Personal Property Assistance: to repair or replace essential household items such as furnishings and appliances, accessibility items (as defined by the Americans with Disabilities Act), and specialized tools and protective clothing required by an employer.
- Transportation Assistance: to repair or replace a vehicle damaged by a disaster and other transportation-related costs.
- Moving and Storage Assistance: to relocate and store personal property from the damaged primary residence to prevent further disaster damage, such as ongoing repairs, and returning the property to the primary residence.

Non-SBA Dependent ONA

Non-SBA dependent types of ONA may be awarded regardless of the individual's or household's SBA disaster loan status and may include the following:

- Funeral Assistance: to assist with funeral expenses incurred as a direct result of a declared major disaster such as reallocation or reburial of unearthed remains and replacement of burial vessel and markers.

---

[10] This information was provided by FEMA congressional liaison to CRS in conversations with the authors. This \$8,500 is the result of average claims that are not covered by insurance or other disaster-related aid.

FEMA and SBA Disaster Assistance for Individuals ...

- Medical and Dental Assistance: to assist with medical or dental expenses caused by a major disaster, which may include injury, illness, loss of prescribed medication and equipment, or insurance copayments.
- Child Care Assistance: a one-time payment that covers up to eight cumulative weeks of child care expenses, for a household's increased financial burden to care for children aged 13 and under, or children aged 14 to 18 with a disability as defined by federal law.
- Miscellaneous or Other Items Assistance: reimbursement for eligible items purchased or rented after a major disaster incident for an individual or household's recovery, such as gaining access to the property or assisting with cleaning efforts.[11]

STAFFORD ACT, SBA DISASTER DECLARATIONS, AND DESIGNATIONS

Two declaration authorities put FEMA IA and the SBA Disaster Loan Program into effect: (1) the Stafford Act, and (2) the Small Business Act (P.L. 85-536, as amended).

Stafford Act Declarations

The Stafford Act authorizes the President to issue major disaster declarations that provide states, tribes, and localities with a range of federal assistance in response to natural and human-caused incidents.[12] Each

---

[11] Federal Emergency Management Agency, *Fact Sheet: Individuals and Households Program*, pp. 2-3, available at https://www.fema.gov/media-library-data/1528984381358-6f256cab09bfcbe6747510c215445560/IndividualsHouseholdsPrograms.pdf.

[12] For more information on major disaster declarations, see CRS Report R43784, *FEMA's Disaster Declaration Process: A Primer*, by Bruce R. Lindsay; and CRS Report R42702, *Stafford Act Declarations 1953-2016: Trends, Analyses, and Implications for Congress*, by Bruce R. Lindsay.

presidential major disaster declaration includes a "designation" listing the counties eligible for assistance as well as the types of assistance FEMA is to provide under the declaration. Potential types of assistance include (1) Public Assistance (PA) for infrastructure repair;[13] (2) Hazard Mitigation Grant Program (HMGP) grants to lessen the effects of future disaster incidents; and (3) Individual Assistance (IA) for aid to individuals and households. Under FEMA regulations:

> The Assistant Administrator for the Disaster Assistance Directorate has been delegated authority to determine and designate the types of assistance to be made available. The initial designations will usually be announced in the declaration. Determinations by the Assistant Administrator for the Disaster Assistance Directorate of the types and extent of FEMA disaster assistance to be provided are based upon findings whether the damage involved and its effects are of such severity and magnitude as to be beyond the response capabilities of the state, the affected local governments, and other potential recipients of supplementary federal assistance. The Assistant Administrator for the Disaster Assistance Directorate may authorize all, or only particular types of, supplementary federal assistance requested by the governor.[14]

Not all major disaster declarations provide IA. Often major declarations only provide PA and HMGP (these are sometimes referred to as "PA-only" major disaster declarations).

Stafford Act disaster declarations also trigger the SBA Disaster Loan Program. The assistance designation, however, determines what loan types become available. In particular, the IA designation is important because it determines whether disaster loans will be made available to individuals and households. For example, if the President declares a major disaster and designates IA for a county, then all SBA disaster loan types become

---

[13]. For more information on PA, see CRS Report R43990, *FEMA's Public Assistance Grant Program: Background and Considerations for Congress*, by Jared T. Brown and Daniel J. Richardson.

[14] 44 C.F.R. §206.40(a).

available to that county.[15] On the other hand, if the President issues a PA-only major declaration, SBA disaster loans are generally only available to private nonprofit organizations. In many cases a major disaster is declared for an incident that designates IA for some counties, and designates PA for others. Only the IA-designated counties in that major disaster will be eligible for SBA home disaster loans.

SBA Disaster Declarations

A major disaster declaration, however, is not the only triggering mechanism for the SBA Disaster Loan Program. The Small Business Act authorizes the SBA Administrator to issue an "Agency" or "SBA declaration" that makes disaster loans available for homeowners, renters, businesses, and nonprofit organizations.[16] The SBA declaration does not, however, trigger FEMA IA.[17]

APPLICATIONS FOR ASSISTANCE

Applying for FEMA Individual Assistance

After a major disaster declaration has been issued and IA has been designated for the incident, applicants in a declared disaster area may

---

[15] Contiguous counties are eligible for Economic Injury Disaster Loans (EIDLs). FEMA assistance is not provided to contiguous counties—only those counties designated in the declaration. The loan types are home disaster loans (Personal Property Loans and Real Property Loans) and business disaster loans (Business Physical Disaster Loans and Economic Injury Disaster Loans).

[16] The criteria used to determine whether to issue a declaration include a minimum amount of uninsured physical damage to buildings, machinery, inventory, homes, and other property. Generally, this minimum is at least 25 homes or businesses (or some combination of the two) that have sustained uninsured losses of 40% or more in any county or other smaller political subdivision of a state or U.S. possession. See 13 C.F.R. §123.3(3)(ii) and 13 C.F.R. §123.3(3)(iii).

[17] Under an SBA declaration, applicants apply directly to SBA for disaster loans. The CMA with FEMA is not used in these situations.

register for FEMA and SBA assistance. Individuals and households can register for assistance online or by telephone.[18] Individuals and households generally have 60 days from the date of a declaration to apply for IA.[19] The registration process typically takes 20 minutes to complete and requires the following information:

- The applicant's social security number (or the social security number of a minor child in the household who is a U.S. citizen, noncitizen national, or qualified alien if the parent or legal guardian is not a legal citizen).
- Financial information (gross household income at the time of the disaster).
- Contract information, insurance information.
- Electronic funds transfer or direct deposit information (to receive eligible assistance).

Applying for an SBA Disaster Loan

Applicants can apply for SBA disaster loans online, in person at a disaster center, or by mail.[20] Applicants must fill out SBA Form 5C and IRS Form 4506-T. The forms require certain information about the applicant including their social security number, income, insurance, assets, debt amounts, and tax information. The applicant may also be required to indicate whether their employment has changed in the last two years, as well as provide deed and proof of residency information. If the applicant is

---

[18] See Appendix C of this report for more information.

[19] Federal Emergency Management Agency, *Individuals and Households Program Unified Guidance*, FP 104-009-03, September 2016, p. 89, available at https://www.fema.gov/media-library-data/1483567080828-1201b6eebf9fbbd7c8a070fddb 308971/FEMAIHPUG_CoverEdit_December2016.pdf.

[20] See Appendix C of this report for more information. See also https://disasterloan.sba.gov/ela/Documents/Three_Step_Process_SBA_Disaster_Loans.pdf.

FEMA and SBA Disaster Assistance for Individuals ... 133

claiming damage to an automobile, they may be required to provide proof of ownership (a copy of the registration, title, bill of sale, etc.).[21]

APPLICANT ELIGIBILITY CRITERIA AND SCREENING

Though integrated to a large extent, ultimately each agency is responsible for determining eligibility based on the applicant's losses and the forms of assistance they have received. In the case of SBA disaster loans, the SBA's Office of Disaster Assistance (ODA) determines eligibility based on the applicant's disaster-related losses, as verified by SBA.[22] IA eligibility determinations are made by FEMA's Individual Assistance Division, under the Office of Response and Recovery.

The SBA Disaster Loan Program uses three main criteria for making credit decisions: (1) eligibility, which is based on the applicant's disaster-related losses; (2) satisfactory credit; and (3) repayment ability, including minimum income levels. SBA will not decline an application for not having collateral to secure a loan but, to the extent it is available, a borrower may be required to pledge collateral for loans over certain amounts (e.g., $25,000 for physical damage loans).[23]

According to FEMA's Individuals and Households Program Unified Guidance, FEMA is required to use four main criteria for determining FEMA IA eligibility: (1) the applicant is a U.S. citizen, noncitizen national, or qualified alien residing; (2) FEMA must be able to verify the applicant's identity; (3) the applicant's insurance, or other forms of disaster assistance received, cannot meet their disaster-caused needs; and (4) the applicant's necessary expenses and serious needs are directly caused by a

---

[21] Puerto Rico residents must also fill out Modelo SC 2907: Release of Inheritance and Donation. See https://www.sba.gov/sites/default/files/articles/f4506-t-2015-09-00.pdf.

[22] Generally based on the amount of disaster damages, minus any assistance received from insurance, FEMA grants, or other sources of recovery.

[23] Small Business Administration, *SBA Disaster Loan Program: Frequently Asked Questions*, August 31, 2017, p. 7, available at https://www.sba.gov/sites/default/files/articles/sba-disaster-loans-faq.pdf.

declared disaster.[24] As described below and discussed in more detail in "SBA Income Test," while FEMA claims that assistance is based on damage amounts rather than income, household income does appear to be a key criterion for making grant determinations.

As demonstrated in Figure 1, applicants typically apply for FEMA IA assistance first. Based on income information provided as part of the FEMA registration process FEMA applies an income test developed by SBA for its initial applicant screening. Applicants above certain income thresholds are not eligible for FEMA IA assistance and are referred back to SBA for loan consideration. SBA developed the minimum income levels or "minimum cost of living estimates" by applying a formula to the U.S Department of Health and Human Services Federal Poverty Guidelines.[25] As shown in Table 1, SBA establishes minimum income levels by multiplying the poverty level for a family of one by a factor of 1.5, and multiplying the poverty level for families of greater than one by a factor of 1.25.[26] The income test formula is not specifically authorized in statute and is not published in SBA regulations. Rather, the formula was first introduced as agency policy in 1985 by SBA Memorandum 85-20 as a means to help determine the applicant's ability to repay the loan and is still in use today (see Appendix B).

FEMA refers applicants to the SBA Disaster Loan Program if they meet or exceed certain income levels and therefore do not qualify for grant assistance. For example, it could be inferred that an individual who earns $18,210 or more per year would initially be denied FEMA grant assistance and referred to the SBA Disaster Loan Program to apply for an SBA disaster loan to repair and rebuild their home (see Stage 2 of Figure 1).

---

[24] U.S. Department of Homeland Security, Federal Emergency Management Agency, *Individuals and Households Program Unified Guidance*, pp. 11-15, https://www.fema.gov/ media-library/assets/documents/124228.

[25] The poverty guidelines are updated on a yearly basis. The U.S Department of Health and Human Services Federal Poverty Guidelines are located at https://aspe.hhs.gov/poverty-guidelines. They are also posted in the *Federal Register*: U.S. Department of Health and Human Services, "Annual Update of the HHS Poverty Guidelines," 83 *Federal Register* 2642-2644, January 18, 2018, available at https://www.federalregister.gov/documents/2018/01/18/2018-00814/annual- update-of-the-hhs-poverty-guidelines.

[26] Minimum income includes wages, alimony, child support payments, interest and dividend income from savings, retirement, pension, social security, or disability payments.

If the applicant is denied an SBA disaster loan, or the loan is insufficient for their recovery needs, the applicant is referred back to FEMA for eligible "SBA dependent ONA" assistance from FEMA (see stage 5 of Figure 1).

SBA disaster loan applicants with income below the minimum income level are classified as Failed Income Test (FIT). FIT applicants are notified that their SBA disaster loan application has been denied and advised that they will be notified if there are any changes to the decision. They are then referred back to FEMA to be considered for possible additional FEMA grant assistance.

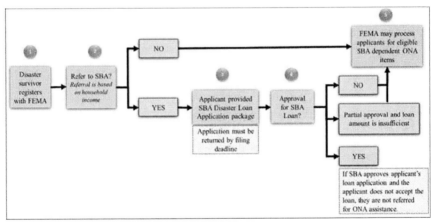

Source: Federal Emergency Management Agency, *Individuals and Households Program Unified Guidance*, FP 104-009-03, September 2016, p. 89, available at https://www.fema.gov/media-library-data/1483567080828-1201b6eebf9fbbd7 c8a070fddb308971/FEMAIHPUG_CoverEdit_December2016.pdf.

Note: ONA= Other Needs Assistance. ONA provides financial assistance for other disaster-related expenses and needs. For more information on ONA see CRS Report R44619, *FEMA Disaster Housing: The Individuals and Households Program—Implementation and Potential Issues for Congress*, by Shawn Reese.

Figure 1. FEMA and SBA Screening Process.

Table 1. SBA Income Test Table
(Poverty Threshold Formula)

Number of Persons in Family/Household	HHS Poverty Guidelines for 2018	SBA Income Threshold
1	$12,140	$18,210
2	$16,460	$20,575
3	$20,780	$25,975
4	$25,100	$31,375
5	$29,420	$36,775
6	$33,740	$42,175
7	$38,060	$47,575
8	$42,380	$52,975

Source: Based on CRS interpretation of U.S Department of Health and Human Services, *U.S. Federal Poverty Guidelines Used to Determine Financial Eligibility for Certain Federal Programs*, Washington, DC, at https://aspe.hhs.gov/poverty-guidelines; and formula applied in SBA Memorandum 85-20: Bernard Kulik, Deputy Associate Administrator for Disaster Assistance, U.S. Small Business Administration, *Income Test Tables:* SBA Memorandum 85-20, June 13, 1985.

FEMA APPEALS

Applicants may submit a written appeal if they disagree with any FEMA determination within 60 days of the date of their eligibility notification letter. Applicants may appeal initial eligibility determinations for housing assistance and ONA, and denials for continued rental assistance, and direct housing assistance. FEMA does not accept multiple appeals for the same reason, but may request additional information and conduct additional reviews as new information is received.

All appeals must be in writing and require an applicant signature—they cannot be accepted via email.[27] The applicant must submit their appeal to the state, territorial, or tribal government if ONA is administered under the Joint or State Option.[28]

---

[27] Appeals must be mailed to FEMA Individuals and Households Program National Processing Service Center, P.O. Box 10055, 20782-8055 or faxed to (800) 827-8112.

[28] Under the joint option, the state, territorial, or tribal government administers ONA jointly with FEMA. FEMA is responsible for registration intake, inspection services, the processing

FEMA and SBA Disaster Assistance for Individuals ... 137

SBA APPEALS

Applicants have six months to request a reconsideration of a decline decision for an SBA disaster loan application. Applicants have 30 days to appeal a subsequent decline decision of their SBA disaster loan application.[29]

POLICY OBSERVATIONS AND CONSIDERATIONS

CMA and Duplication of Benefits

Following a major disaster, homeowners and businesses may have access to a number of resources to assist in the response, recovery, and rebuilding process. The range of resources includes insurance payouts, state and local government assistance, charitable donations from private institutions and individuals, as well as certain forms of federal assistance. In addition to FEMA and SBA disaster assistance, individuals and households may be eligible for the Department of Housing and Urban Development's (HUD's) Community Development Block Grant Disaster Recovery (CDBG-DR) program.[30] Compensation from multiple sources that exceed the total loss amount is generally considered a duplication of

---

system, mail processing, and accessible forms of communication. The state, territorial, or tribal government is responsible for manually processing awards, staffing, recovery of funds, case processing, appeals, and preparing closeout material. Under the State Option, FEMA provides ONA as a grant to the state, territorial, or tribal government; therefore, the state, territorial, or tribal government administers ONA. The state, territorial, or tribal government is responsible for all tasks associated with the administration of ONA. See Federal Emergency Management Agency, *Individuals and Households Program Unified Guidance*, FP 104-009-03, September 2016, p. 91, available at https://www.fema.gov/media-library-data/1483567080828-1201b6eebf9fbbd7c8a070fddb 308971/FEMAIHPUG_CoverEdit_December2016.pdf.

[29] 13 C.F.R. §123.13(e).

[30] For more information on the Community Development Block Program, see CRS Report R43520, *Community Development Block Grants and Related Programs: A Primer*, by Eugene Boyd; CRS Report RL33330, *Community Development Block Grant Funds in Disaster Relief and Recovery*, by Eugene Boyd; and CRS Report R43394, *Community Development Block Grants: Recent Funding History*, by Eugene Boyd.

benefits. When duplication occurs, the recipient is liable to the United States to pay back the duplicated benefit.

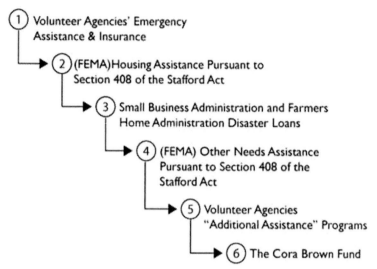

Source: CRS interpretation of 44 C.F.R. 206.191.

Note: Housing assistance under Section 408 includes assistance to individuals and households who are displaced from their predisaster primary residences or whose predisaster primary residences are rendered uninhabitable, or, with respect to individuals with disabilities, rendered inaccessible or uninhabitable as a result of a major disaster. Section 408 includes temporary housing as well as repairs. Other needs assistance under Section 408 includes financial assistance for medical, dental, and funeral expenses. Cora Brown of Kansas City Missouri died in 1977. She left a portion of her estate to the United States to be used as a special fund to relieve human suffering caused by natural disasters. For more information on the Cora Brown Fund see https://www.fema.gov/ media-library-data/1434639028239-341e17807cb06b0bf000d21cc7552b2c/CoraBrown-FactSheet-final508.pdf.

Figure 2. Disaster Assistance Delivery Sequence.

Section 312 of the Stafford Act requires federal agencies to ensure that individuals (and businesses) do not receive disaster assistance for losses for which they have already been compensated or may expect to be compensated.[31] FEMA is the primary agency responsible for policy and procedural guidance on duplication of benefits. The uniformity requirement set forth in Section 312 of the Stafford Act is located in FEMA regulation 44 C.F.R. §206.191, which establishes a delivery sequence of disaster assistance provided by federal agencies and organizations (see Figure 2).

An organization's position within the sequence determines the order in which it should provide assistance and what other resources must be considered before that assistance is provided. Further, each organization is responsible for delivering assistance without regard to duplication later in the sequence.

According to FEMA regulations, the agency or organization that is lower in the delivery sequence should not provide assistance that duplicates assistance provided by an agency or organization higher in the sequence. When the delivery sequence has been disrupted, the disrupting agency is responsible for rectifying the duplication.

As mentioned previously in this chapter, SBA and FEMA have a computer matching agreement (CMA) to share real-time data on assistance provided to applicants. SBA also shares relevant data with states, territories, tribes, and local government jurisdictions and voluntary agencies. In cases where the data contain personally identifiable information (PII), a Memorandum of Understanding (MOU) must first be prepared and signed by both SBA and the authorized requesting party to help avoid duplication of benefits.[32] SBA and FEMA entered into the CMA pursuant to Section (o) of the Privacy Act of 1974 (5 U.S.C. §552a).[33]

---

[31] 42 U.S.C. §5155. There are numerous statutes and regulations that prohibit duplication of benefits with respect to disaster assistance. These are included in Appendix A.

[32] The MOU template can be located at https://es.sba.gov/sites/default/files/articles/sba_disaster_data_sharing_mou_and_instructions_0.docx.

[33] As amended by the Computer Matching and Privacy Protection Act of 1988 (P.L. 100-503), and as amended by the Computer Matching and Privacy Protection Amendments of 1990 (P.L. 101-508, 5 U.S.C. §552a(p) (1990)).

As outlined in the September 25, 2015, *Federal Register* notice, "the financial and administrative responsibilities will be evenly distributed between SBA and DHS/FEMA unless otherwise set forth in this agreement."[34] The *Federal Register* notice further stated that the CMA is "part of a government-wide initiative, Executive Order 13411—Improving Assistance for Disaster Victims (August 29, 2006) ... to ensure that benefits provided to disaster survivors by DHS/FEMA and SBA are not duplicated."[35]

Though the CMA is part of a government-wide initiative to prevent duplication of benefits, it is solely used by FEMA and SBA. Some may argue that the CMA should also be used by other federal agencies that offer disaster assistance (such as CDBG-DR). Others, however, might argue using the CMA with agencies other than FEMA and SBA is problematic or potentially ineffective. For example, HUD provides funds to state and local jurisdiction grantees. HUD does not have the individual award data to match up with SBA loan applicant data to determine instances of duplication.

If Congress is concerned about the use of the CMA, it could investigate how it could be used in conjunction with other disaster assistance programs. If CMA is not being universally used by federal agencies, Congress could investigate why it is not being used, evaluate challenges preventing its use, and mandate its use across the federal government. Additionally, Congress could use oversight to investigate how effective the CMA has been at preventing duplication since its implementation.

Similarly, Congress could also review how MOUs are being used by states, territories, tribes, local government jurisdictions, and voluntary agencies to determine their efficacy in reducing duplication.

---

[34] Small Business Administration/Federal Emergency Management Agency, "Privacy Act; Computer Matching Agreement," 80 *Federal Register* 57902-57906, September 25, 2015.
[35] Ibid.

FEMA and SBA Disaster Assistance for Individuals ... 141

Congress could also consider making the MOUs a requirement for certain types of assistance or investigate other incentives that would induce the entities to sign MOUs.

SBA Income Test

As mentioned previously, both FEMA and SBA screen applicants according to minimum income levels (see Table 1). SBA creates the minimum income levels by applying a formula to the U.S Department of Health and Human Services Federal Poverty Guidelines. According to SBA Standard Operation Procedure (SOP), minimum income includes wages, alimony, child support payments, interest and dividend income from savings, retirement, pension, social security, or disability payments.

According to SBA Memorandum 85-20 (see Appendix B), the income test table is used as a guide "for summary declines." It is unclear from SBA SOPs if income levels are a hard limit to screen out applicants, or if there is some discretion in the process. Additionally, some might be confused about income standards based on publicly available information provided to disaster survivors. For example, in a news release published in the wake of the 2011 Missouri River flooding, FEMA stated "federal and state disaster assistance programs are available to all who suffered damages. Aid is damage-based, not income-based. The kinds of help provided depend on an applicant's circumstances and unmet needs."[36] In another news release issued May 9, 2014, FEMA stated that "income is not a consideration for FEMA disaster assistance."[37]

---

[36] Federal Emergency Management Agency, *Commonly Asked Questions About Disaster Aid*, November 15, 2011, available at https://www.fema.gov/news-release/2011/11/15/commonly-asked-questions-about-disaster-aid.

[37] Federal Emergency Management Agency, *Myth vs. Fact: The Truth About Registering for Federal Disaster Assistance*, May 9, 2014, available at https://www.fema.gov/news-release/2014/05/09/myth-vs-fact-truth-about-registering-federal-disaster-assistance.

On the one hand, it could be argued that the income test is being used by FEMA as a "pre-decisional" screening tool, and that ultimately, income is not a factor for FEMA grant assistance. Others may disagree and question why FEMA reviews household income if it has no bearing on grant assistance. If Congress does not want grant determinations to be based on income, it could prohibit FEMA from collecting income information and using it as a screening tool. If, however, Congress determines that grant determinations should be based, at least in part, on certain income levels, it could require FEMA to publish policy documents that specify how income is used to award grant assistance.

Congress may also consider whether SBA should continue to use established thresholds and formulas based on SBA Memorandum 85-20 to determine eligibility, or provide SBA with some measure of discretion in the process. Some might argue that uniformity is needed to ensure equitable determinations. Others might argue that a one size-fits-all approach does not address special or mitigating circumstances. In general, eligibility for an SBA disaster loan is assigned to the person (or entity in the case of businesses) that legally owns or is responsible for the repair or replacement of the disaster-damaged property based on that person's ability to repay the loan. It could be argued, however, that the ability to pay a loan is not solely determined by income and that a range of factors and circumstances should also be considered. For example, a retired person may not meet a certain minimum income level, but own assets that could be liquidated for repayment purposes. Another example is the parent(s) of a university student who are willing to cosign for their child's disaster loan. In both these examples, strict adherence to an income test might prevent people from obtaining loans who may be able to repay them through nontraditional methods. Congress may also consider whether the income test is an ineffective screening tool for identifying applicants that meet the income test but cannot repay their disaster loan; for example, a person who earns more than the minimum income level, but has debt that impedes their ability to repay the disaster loan.

FEMA and SBA Disaster Assistance for Individuals ... 143

Finally, it is not clear if FEMA has access to the same income information as SBA. As mentioned earlier, minimum income includes wages, alimony, child support payments, interest and dividend income from savings, retirement, pension, social security, and disability payments. This information is gathered on SBA Form 5C and IRS Form 4506-T. Applicants, however, can fill these forms out after initially registering with FEMA. This could result in the reporting of different income levels. For example, the income reported to FEMA could be less than what is reported to SBA.

If Congress is concerned about how determinations are made, it could consider requiring SBA and FEMA to publish specific determination criteria in their respective regulations and policy guidance documents. Congress may also consider putting the determination formula into statute. Finally, Congress may decide whether it is best to have uniformity in the determination process, or whether to provide the agencies certain parameters that allow for some discretion when making grant and loan decisions.

APPENDIX A. RELEVANT STATUTORY AUTHORITIES AND REGULATIONS

The following is a listing of selected authorities and regulations pertaining to the duplication of disaster assistance benefits. This list should be considered representative, not exhaustive.

Stafford Act (42 U.S.C. §5155)

The Stafford Act is the primary statute governing the provision of federal disaster assistance, particularly FEMA assistance. Section 312 of the Stafford Act requires federal agencies that provide financial disaster assistance to ensure that individuals, businesses, or other entities suffering losses as a result of a major disaster or emergency do not receive assistance for losses for which they have already been compensated. Section 312 also requires the President to establish procedures that ensure uniformity in preventing duplication of benefits. Under Section 312, any person, business, or other entity that has received or is entitled to receive federal disaster assistance is liable to the United States for the repayment of such assistance to the extent that such assistance duplicates benefits available for the same purpose from another source, including insurance and other federal programs.

Stafford Act (42 U.S.C. §5174)

Section 408(a)(1) states that the President may provide assistance to individuals and households who, as a result of a major disaster, "have necessary expenses and serious needs in cases in which the individuals and households are unable to meet such expenses or needs through other means."

FEMA Regulation

44 C.F.R. 206.191, which establishes the policies implementing Section 312 of the Stafford Act, states that it is FEMA policy to prevent the duplication of benefits between its own programs, other assistance programs, and insurance benefits. The regulation requires individuals to repay all duplicated assistance to the agency providing the assistance. Under 44 C.F.R. 206.191, a federal agency providing disaster assistance is responsible for preventing or rectifying duplication of benefits when they occur. 44 C.F.R. 206.191 also includes a "delivery sequence" hierarchy intended to prevent waste, fraud, and abuse of program assistance, including the duplication of benefits (see Figure 2).

44 C.F.R. 206.111 defines financial ability of the applicant to pay housing costs. According to 44 C.F.R. 206.111 if the "household income

FEMA and SBA Disaster Assistance for Individuals ... 145

has not changed subsequent to or as a result of the disaster then the determination is based upon the amount paid for housing before the disaster. If the household income is reduced as a result of the disaster then the applicant will be deemed capable of paying 30 percent of gross post disaster income for housing. When computing financial ability, extreme or unusual financial circumstances may be considered by the Regional Administrator."

Small Business Act (15 U.S.C. §636(b)(1)(A))

The first proviso in 15 U.S.C. §636(b)(1)(A) states that SBA is authorized to make disaster loans "provided that such damage or destruction is not compensated for by insurance or otherwise."

Small Business Act (15 U.S.C. §647)

Section 18(a) of the Small Business Act (P.L. 85-536, as amended) prohibits the SBA from providing benefits that duplicate the assistance provided by another department or agency of the federal government. Section 18(a) states that if loan applications are refused or denied by a department or agency due to administrative withholding or due to an administratively declared moratorium, then no duplication is deemed to have occurred.

SBA Regulation

13 C.F.R. 123.101(c) states that applicants for SBA Disaster Loan assistance are not eligible for a home disaster loan if their damaged property can be repaired or replaced with the proceeds of insurance, gifts, or other compensation. These amounts must either be deducted from the

146 *Bruce R. Lindsay and Shawn Reese*

amount of the claimed losses or, if received after SBA has disbursed the loan, must be paid to SBA as principal payments on the loan.

APPENDIX B. SBA MEMORANDUM 85-20

Date: JUN 13 1985

To: All Area Directors # 85-20

From: Bernard Kulik, DAA/DA

Subject: Income Test Tables

Enclosed you will find a set of income test tables to be used as guides for summary declines. The information is from Health and Human Services (HHS) 1985 Poverty Income Guidelines. We increased the single member family unit by 150 percent and all other units by 125 percent to arrive at the minimum cost of living estimates. The Puerto Rico table is for disaster declarations on or after May 31, 1985, and the other tables for disaster declarations on or after June 10, 1985. No other income guides are acceptable for use as minimum cost of living estimates.

(Signed) Bernard Kulik

Bernard Kulik
Deputy Associate Administrator
for Disaster Assistance

Enclosures

DAD:Allred:JJ:6/11/85:0906d

cc: Subject   Reading   Central Files   Allred

Figure B-1. SBA Memorandum 85-20.

In: Issues in Disaster Recovery and Assistance   ISBN: 978-1-53616-308-7
Editor: Donatien Moïse   © 2019 Nova Science Publishers, Inc.

Chapter 5

DISASTER ASSISTANCE: FEMA ACTION NEEDED TO BETTER SUPPORT INDIVIDUALS WHO ARE OLDER OR HAVE DISABILITIES*

United States Government Accountability Office

ABBREVIATIONS

ADA	Americans with Disabilities Act of 1990, as amended
CRCL	Department of Homeland Security Office for Civil Rights and Civil Liberties
DIA	Disability Integration Advisors
DHS	Department of Homeland Security
FEMA	Federal Emergency Management Agency
IA	FEMA's Individual Assistance program
IHP	FEMA's Individuals and Households Program

---

* This is an edited, reformatted and augmented version of United States Government Accountability Office; Report to Congressional Requesters, Publication No. GAO-19-318, dated May 2019.

NRF	National Response Framework
ODIC	FEMA's Office of Disability Integration and Coordination
OMB	Office of Management and Budget
Post-Katrina Act	Post-Katrina Emergency Management Reform Act of 2006
Rehabilitation Act	Rehabilitation Act of 1973, as amended
RDIS	Regional Disability Integration Specialist
Stafford Act	Robert T. Stafford Disaster Relief and Emergency Assistance Act, as amended
TTY	Text Telephone

WHY GAO DID THIS STUDY

Three sequential hurricanes—Harvey, Irma, and Maria—affected more than 28 million people in 2017, according to FEMA. Hurricane survivors aged 65 and older and those with disabilities faced particular challenges evacuating to safe shelter, accessing medicine, and obtaining recovery assistance. In June 2018, FEMA began implementing a new approach to assist individuals with disabilities.

GAO was asked to review disaster assistance for individuals who are older or have disabilities. This chapter addresses (1) challenges FEMA partners reported in providing assistance to such individuals, (2) challenges such individuals faced accessing assistance from FEMA and actions FEMA took to address these challenges, and (3) the extent to which FEMA has implemented its new approach to disability integration.

GAO analyzed FEMA data and reviewed relevant federal laws, agency policy, and federal frameworks. GAO also interviewed state, territorial, local, and nonprofit officials in Florida, Puerto Rico, Texas, and the U.S. Virgin Islands; FEMA officials at headquarters, in regional offices, and deployed to disaster sites; and officials at relevant nonprofit organizations.

WHAT GAO RECOMMENDS

GAO is making seven recommendations to FEMA, including that it establish new registration questions, objectives for its new disability integration approach, and a training plan for FEMA staff. The agency concurred with all but one of the recommendations.

WHAT GAO FOUND

A range of officials from entities that partner with the Federal Emergency Management Agency (FEMA)—including states, territories, localities, and nonprofits)—reported challenges providing assistance to individuals who are older or have disabilities following the 2017 hurricanes. For example, officials said that many of these individuals required specialized assistance obtaining food, water, medicine, and oxygen, but aid was sometimes difficult to provide. Officials in Puerto Rico and the U.S. Virgin Islands cited particular difficulties providing this assistance due to damaged roads and communication systems, as well as a lack of documentation of nursing home locations.

Based on GAO's analysis of FEMA data and interviews with FEMA officials and stakeholders, aspects of the process to apply for assistance from FEMA after the 2017 hurricanes were challenging for older individuals and those with disabilities. According to stakeholders and FEMA officials, disability-related questions in the registration materials are confusing and easily misinterpreted. For example, FEMA's registration process does not include an initial question that directly asks individuals if they have a disability or if they would like to request an accommodation for completing the application process (see figure below). While FEMA has made efforts to help registrants interpret the questions, it has not yet changed the language of the questions to improve clarity. As a result, individuals with disabilities may not have requested accommodations or reported having disabilities, which may have hindered FEMA's ability to identify and assist them.

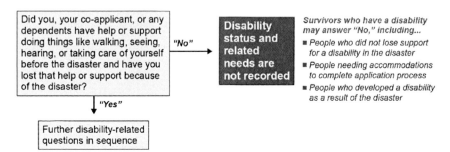

Source: GAO analysis of Federal Emergency Management Agency (FEMA) documentation. | GAO-19-318.

Sequence of Disability-Related Questions in FEMA's Registration Process.

FEMA did not establish objectives before implementing its new approach to disability integration, which includes adding new disability integration staff in the regions and decreasing the number of disability integration advisors deployed to disaster sites. Without documented objectives for the new approach, regional leadership across the nation may implement changes inconsistently. In addition, the new approach shifts the responsibility for directly assisting individuals with disabilities to all FEMA staff. FEMA has taken some initial steps to provide training on the changes; however, it has not established a plan for delivering comprehensive disability-related training to all staff who will be directly interacting with individuals with disabilities. Developing a plan to train all staff would better position FEMA to achieve its intended goals and better equip deployed staff to identify and assist these survivors.

May 14, 2019

Congressional Requesters

In the summer and fall of 2017, Hurricanes Harvey, Irma, and Maria affected more than 28 million people living in Texas, the U.S. Virgin Islands, Florida, and Puerto Rico and caused a combined $265 billion in

Disaster Assistance 151

damage, according to the Federal Emergency Management Agency (FEMA), a component of the Department of Homeland Security. As we previously reported, the consecutive timing and scale of the disasters overwhelmed the capabilities of some local, state, and territorial governments, as well as nonprofit partners, to assist affected residents.[1]

Individuals affected by hurricanes and other large-scale disasters who are older or have disabilities can face particular challenges obtaining disaster assistance.[2] Some individuals who are older or have disabilities, and who otherwise function independently in their day-to-day lives, may rely on supports that disasters can interrupt. For example, the two suppliers of oxygen on the island of Puerto Rico lost production capabilities after Hurricane Maria due to a lack of power, which in turn threatened the health of the reported 50,000 Puerto Ricans who depended on oxygen, according to a disability rights organization's report.

Emergency management and private organization partners turned to FEMA for help, including from FEMA disability integration staff who were responsible for providing assistance to individuals with disabilities regardless of age.[3] However, the hurricanes also overwhelmed FEMA's

---

[1] GAO, *2017 Hurricanes and Wildfires: Initial Observations on the Federal Response and Key Recovery Challenges,* GAO-18-472 (Washington, D.C.: Sept.4, 2018). Because sections of this report focus on the impact of the 2017 hurricanes on populations in specific locations, we use the phrase "state, territorial, and local" to refer to the state, county, municipal, and territorial governments in the affected areas and the officials who run them. In generally referring to emergency management partners of the Federal Emergency Management Agency (FEMA) at nonfederal levels of government, we use the phrase "state and local." In this latter context, the phrase includes state, tribal, territorial, insular area, and local governments as well as the government of the District of Columbia.

[2] For the purposes of this report, the phrase "individuals with disabilities" refers to individual disaster survivors, including those who are 65 or older, who have a disability that affects their ability to evacuate, shelter, or recover from a disaster. Under federal civil rights laws, an individual with a disability is generally defined as an individual who has a physical or mental impairment that substantially limits one or more major life activities. FEMA provides specialized services to those with "access and functional needs," which includes, among others, individuals with disabilities, older adults, and individuals with limited English proficiency, limited access to transportation, and/or limited access to financial resources to prepare for, respond to, and recover from a disaster. For the purposes of this report, we use the term "individuals who are older" to refer to individuals who are age 65 or older, regardless of whether they have a disability or an access or functional need.

[3] Disability integration staff are responsible for focusing on inclusive practices in emergency management, and include those deployed to areas affected by disasters and those working permanently in FEMA's regional offices.

152 *United States Government Accountability Office*

available workforce, and the agency reported that it met approximately 50 percent of its target for deploying disability integration staff that could be deployed to the sites of the 2017 hurricanes and less than half of its target for the number of disability integration staff who were qualified.[4]

After the 2017 hurricane season, FEMA announced plans to reorganize its workforce to more thoroughly incorporate disability integration principles into all preparedness, response, and recovery activities nationwide and reduce reliance on FEMA's disability integration staff. This reorganization was announced to begin in June 2018, near the start of the 2018 hurricane season.

You asked us to review the federal government's response to the 2017 hurricanes with a focus on older adults and people with disabilities. This chapter is also one in a series that addresses the federal response to those disasters.

This chapter examines (1) challenges FEMA's partners reported in providing disaster assistance to individuals who are older or have disabilities; (2) challenges faced by these individuals in accessing FEMA's disaster assistance programs and actions FEMA has taken to address such challenges; and (3) the extent to which FEMA has implemented its new approach to disability integration.

To address our first and second objectives, we visited Florida, Puerto Rico, Texas, and the U.S. Virgin Islands in June and July 2018.[5] At each location we interviewed state or territory emergency managers, public health and human services officials, and representatives of nonprofit disability organizations. For example, we interviewed staff from Centers for Independent Living in each location. To learn first-hand accounts of disaster-related challenges faced by individuals in Puerto Rico and the U.S. Virgin Islands who are older or have disabilities and who were impacted by the hurricanes, staff from the Centers for Independent Living in those locations invited us to interview 16 of their regular program participants. We interviewed local emergency managers in Texas and Florida in

---

[4] FEMA, *2017 Hurricane Season FEMA After-Action Report* (Washington, D.C.: July 12, 2018).

[5] Hurricane Harvey primarily affected the Gulf Coast of Texas; Hurricane Irma primarily affected the U.S. Virgin Islands, Puerto Rico, and Florida; and Hurricane Maria primarily affected the U.S. Virgin Islands and Puerto Rico.

counties that were affected by Hurricanes Harvey and Irma. We also interviewed representatives of national organizations, selected for their focus on providing disaster assistance to individuals who are older or have disabilities. In addition, we interviewed officials from other relevant federal agencies and offices, including the Department of Homeland Security's (DHS) Office for Civil Rights and Civil Liberties (CRCL) and the National Council on Disability. To supplement information we obtained from the site visit interviews, we reviewed summaries of eight public listening sessions published by CRCL and co-hosted with FEMA across the four disaster locations between February 2018 and May 2018. While the perspectives of officials and stakeholders we interviewed, as well as those expressed during the public listening sessions, are not generalizable, they provide valuable insights into the federal response to the 2017 disasters.

To address our second objective regarding the disaster assistance FEMA provides, we obtained and analyzed summary data from FEMA's primary database on its registrations and awards for survivors of Hurricanes Harvey, Irma, and Maria. FEMA also provided disaster assistance data on registrations submitted by households with residents who are older and households with residents reporting disabilities. We also obtained and analyzed data from call centers that operate FEMA's helpline, including the number of incoming calls and the average wait time for answered calls for a given day. To assess the reliability of FEMA's data, we interviewed officials at FEMA headquarters about the quality of the data; reviewed existing information about the data systems; and conducted checks for inaccurate data and comparisons to publicly available summary data. We determined that the data we obtained were sufficiently reliable for the purposes of providing information on the number and characteristics of registrations for assistance FEMA collects and the number of calls to FEMA's call centers.

To address our third objective regarding FEMA's new approach to disability integration, we compared staffing levels before and after FEMA's workforce reorganization by obtaining and analyzing the number of disability integration staff deployed in response to the 2017 hurricanes

154 *United States Government Accountability Office*

and to Hurricanes Florence and Michael, which made landfall in September and October 2018. To assess the reliability of these data, we reviewed recent GAO work that assessed the reliability of FEMA workforce data from the same data source and reviewed the data for obvious errors and omissions.[6] We determined that the data we obtained were sufficiently reliable for the purposes of providing information on the deployed FEMA workforce in response to recent hurricanes. We also obtained and analyzed responses to structured questions about FEMA's new approach to disability integration from officials in FEMA's ten regions.

To address all three objectives, we assessed FEMA's efforts to assist individuals who are older or have disabilities against relevant criteria. We analyzed FEMA policies, procedures, guidance, and memoranda, including those specific to FEMA's Individuals and Households Program and disability integration. For example, we analyzed memoranda describing FEMA's plan to implement its new approach to disability integration and the agency's policy on data-sharing. We also reviewed FEMA's internal self-evaluation of its policies and practices, which it conducted to evaluate how effectively FEMA provides equal physical, program, and communication access to people with disabilities.[7] We assessed these documents against goals and objectives in FEMA's 2018-2022 Strategic Plan, DHS policy for ensuring nondiscrimination for individuals with disabilities, and federal standards for internal control related to communicating effectively internally and externally, using quality information to achieve objectives, and defining objectives in measurable terms.[8] We reviewed relevant information from our prior reports on

---

[6] GAO-18-472.

[7] FEMA completed its self-evaluation in August 2017 and officials told us they have plans to publish a plan for addressing issues identified in the self-assessment in 2019. GAO reviewed the self-evaluation and a draft implementation plan in December 2018.

[8] See FEMA, *2018-2022 Strategic Plan*, March 15, 2018; DHS Directive 065-01, Nondiscrimination for Individuals with Disabilities in DHS-Conducted Programs and Activities (Non-Employment), Sept. 25, 2013; GAO, *Standards for Internal Control in the Federal Government*, GAO-14-704G (Washington, D.C.: September 2014).

Disaster Assistance 155

FEMA's work,[9] as well as The Partnership for Inclusive Disaster Strategies' 2018 After Action Report.[10] We did not independently assess whether any programs or activities conducted by FEMA or its partners during the period covered by our review complied with applicable non-discrimination or civil rights laws.

We also interviewed FEMA officials from headquarters and staff deployed to each disaster location, including staff focused on assisting individuals with disabilities. We interviewed former FEMA officials, including a previous FEMA administrator, for historical perspective on the changes FEMA is making to disability integration.

We conducted this performance audit from April 2018 to May 2019, in accordance with generally accepted government auditing standards. Those standards require that we plan and perform the audit to obtain sufficient, appropriate evidence to provide a reasonable basis for our findings and conclusions based on our audit objectives. We believe that the evidence obtained provides a reasonable basis for our findings and conclusions based on our audit objectives.

Further information on our scope and methodology can be found in appendix I.

BACKGROUND

Aging and Disability in Disasters

Older individuals and individuals with disabilities, whose needs may overlap, may face particular risks in disasters, including hurricanes.

---

[9] GAO-18-472; *Federal Disaster Assistance: FEMA's Progress in Aiding Individuals with Disabilities Could Be Further Enhanced*. GAO-17-200, (Washington, D.C.: Feb. 7, 2017); and *Actions Taken to Implement the Post-Katrina Emergency Management Reform Act of 2006*, GAO-09-59R (Washington, D.C.: Nov. 21, 2008).

[10] This report incorporated the perspectives of stakeholders who were working in disaster-impacted communities and thousands of callers to a hotline managed by The Partnership to address the needs of survivors of the 2017 hurricanes. The Partnership for Inclusive Disaster Strategies, *Getting It Wrong: An Indictment with a Blueprint for Getting It Right. Disability Rights, Obligations and Responsibilities Before, During and After Disasters* (May 2018).

156 *United States Government Accountability Office*

According to AARP, older persons may be disproportionately vulnerable to disasters because they are more likely to have chronic illnesses, functional limitations, and sensory, physical, and cognitive disabilities than those who are younger.[11] Following Hurricane Katrina in 2005, the National Council on Disability noted that the basic needs of people with disabilities were compounded by chronic health conditions and functional impairments, including blindness, hearing deficiencies, mobility impairments, and mental health conditions.[12] As a result of chronic health conditions and functional limitations, older persons may often take multiple medications, rely on caregivers for assistance, and experience general "frailty." In particular, individuals who are frail or who have disabilities and who live alone or in isolated rural areas may be more vulnerable to the effects of disasters like hurricanes.

According to 2017 Census estimates, 15.4 percent of the U.S. population not living in an institution is 65 or older, and 12.7 percent has a disability. More than one-third (34.6 percent) aged 65 or older has a disability, and 42.1 percent of those with disabilities are 65 or older.[13]

State and Local Disaster Assistance

State and local governments are primarily responsible for disaster management, but the federal government, as well as the nonprofit and business sectors, can provide critical support if those entities need assistance. The National Response Framework (NRF) defines the roles of entities that respond to all types of incidents, such as hurricanes and other disasters, including local and state governments, the federal government, the private sector, and voluntary organizations. One of the NRF's core

---

[11] Mary Jo Gibson, *We Can Do Better: Lessons Learned for Protecting Older Persons in Disasters* (Washington, D.C.: AARP, 2006).

[12] National Council on Disability, *The Impact of Hurricanes Katrina and Rita on People with Disabilities: A Look Back and Remaining Challenges* (Washington, D.C.: Aug. 3, 2006).

[13] Data were obtained from the 2017 American Community Survey. Estimates have a margin of error at the 90 percent confidence interval of plus or minus 0.1 percentage points.

principles is that response efforts must adapt to meet evolving demands resulting from changes in disaster size, scope, and complexity.

Under the NRF, state and local agencies are primarily responsible for response and recovery activities in their jurisdictions, including those involving health and safety. For example, state and local agencies are primarily responsible for carrying out evacuations and administering shelters, when necessary, for those affected by a disaster. In addition, the NRF emphasizes the importance of state and local emergency management agencies coordinating disaster assistance with the private sector. In particular, organizations composing states' and localities' critical infrastructure—such as private hospitals—may serve as partners in disaster preparedness and response. As part of their responsibilities, state and local governments provide financial and nonfinancial disaster assistance, and work alongside FEMA staff and voluntary organizations to connect disaster survivors with outside resources. Certain types of assistance, such as evacuations and sheltering, can happen before the disaster and are particularly important for people who are older or have disabilities. Some states, including Texas and Florida, established voluntary registries for residents who may need specialized assistance in an emergency. In states that offer a registry, residents who depend on a consistent source of electricity to power medical equipment, for instance, might opt to register as a way of communicating their need for access to generator power in the case of a widespread power outage.

Federal Disaster Assistance

When needs for assistance resulting from a disaster exceed or are expected to exceed state or local resources, the federal government may use the NRF to involve all necessary federal department and agency capabilities and ensure coordination with response partners. The Robert T. Stafford Disaster Relief and Emergency Assistance Act, as amended

158 *United States Government Accountability Office*

(Stafford Act)[14] outlines the process state and local governments can use to obtain federal support under the act in response to a disaster. First, a state's governor must submit a request to the President to declare a major disaster in the state.[15] If the President grants the declaration, the state becomes eligible for various types of assistance from FEMA, such as personnel, funding, and technical assistance, among others. In the case of a federally declared disaster under the Stafford Act, FEMA has primary responsibility for coordinating the federal response, and it targets the level of federal support to the needs specified by states' requests for assistance. FEMA staff deployed to disasters work alongside state counterparts and voluntary organizations at a centralized location, called the joint field office, to coordinate disaster response and recovery efforts.[16]

One of the main forms of federal assistance comes through FEMA's Individuals and Households Program (IHP), one of six subprograms under FEMA's Individual Assistance (IA) program.[17] IHP provides financial and direct assistance to eligible individuals and households who have

---

[14] See generally 42 U.S.C. § 5121 *et seq.*; 44 C.F.R. pt. 206.

[15] The governor's request shall be based on a finding that the disaster is of such severity and magnitude that effective response is beyond the capabilities of the state and the affected local governments and that federal assistance is necessary. A major disaster is defined as any natural catastrophe (e.g., a hurricane, tornado, snowstorm, or earthquake) or, regardless of cause, any fire, flood, or explosion, in any part of the U.S., which in the determination of the President causes damage of sufficient severity and magnitude to warrant major disaster assistance under the act to supplement the efforts and available resources of states, local governments, and disaster relief organizations. An emergency is defined as any occasion or instance for which, in the determination of the President, federal assistance is needed to supplement state and local efforts and capabilities to save lives and to protect property and public health and safety, or to lessen or avert the threat of a catastrophe in any part of the U.S. 42 U.S.C. §§ 5170, 5122; 44 C.F.R. §§ 206.31- 206.48. For purposes of this report, we focus only on presidentially-declared major disasters.

[16] Federal leadership at the joint field office includes the Federal Coordinating Officer and the Chief of Staff, among others.

[17] In addition to Individual Assistance, FEMA also provides disaster recovery funding assistance through its Public Assistance and Hazard Mitigation Grant programs. The Public Assistance program primarily provides supplemental federal disaster grant assistance to state, local, tribal, and territorial governments, and certain types of private nonprofit organizations for debris removal, emergency protective measures, and the restoration of disaster-damaged, publicly-owned facilities and the facilities of certain private nonprofit organizations. The Hazard Mitigation Grant Program is designed to improve community resilience and funds a wide range of projects, such as purchasing properties in flood-prone areas, adding shutters to windows to prevent future damage from hurricane winds and rains, and modifying culverts in drainage ditches to prevent future flooding damage.

Disaster Assistance 159

uninsured or underinsured necessary expenses and serious needs.[18] Under IHP, individuals can receive two types of assistance: housing assistance and other needs assistance.

- Housing assistance can come in the form of financial or direct assistance for temporary housing, home repairs, replacement of a primary home, or in limited locations, permanent housing construction when needed due to disaster effects.[19] To receive housing assistance, individuals typically must participate in a home inspection to verify certain information required to determine eligibility, including damage and loss.

- Other needs assistance provides financial assistance to replace or repair uninsured or underinsured personal property, or support disaster-related needs, such as transportation, funeral, medical, and child care assistance.

To receive FEMA assistance under IHP, individuals must register by answering a standard series of registration intake questions. The registration intake process is designed to solicit relevant information from individuals to determine their eligibility for certain FEMA disaster assistance programs.[20] The process collects some basic demographic information, including name, age, number of people living in the household, and how they were affected by the disaster. It also requires individuals to report financial information, such as insurance status, which may affect eligibility for certain programs. Individuals can register by phone using a toll-free helpline, which includes Text Telephone (TTY) and Video Relay Service capabilities for individuals who are deaf, hard of hearing, or have a speech disability; via the internet or a smartphone; or in

---

[18] See 42 U.S.C. § 5174; 44 C.F.R. §§ 206.101-206.120.

[19] Permanent housing construction assistance may be available in insular areas outside the continental United States (such as the U.S. Virgin Islands) and other locations in which alternative housing resources are not available and other temporary housing assistance is unavailable, infeasible, or not cost-effective.

[20] Based on applicants' registration responses, FEMA may also provide referrals to other agencies or nonprofits.

160 *United States Government Accountability Office*

person at Disaster Recovery Centers. According to FEMA, Disaster Recovery Centers are usually opened quickly after a disaster for a limited period of time, are equipped to accommodate those who need disability-related communication aids, and are established in coordination with state and local governments. Also according to FEMA, staff are available to assist individuals registering for assistance through the helpline, at Disaster Recovery Centers, and in their communities.

Legal Protections for Older Individuals and Those with Disabilities and Inclusive Emergency Management Practices

In emergency management, disability integration includes incorporating inclusive practices and applicable requirements related to individuals with disabilities—such as those that may apply under the Rehabilitation Act of 1973 (Rehabilitation Act) or the Americans with Disabilities Act of 1990 (ADA), both as amended—into all aspects of emergency preparedness and disaster response, recovery, and mitigation.[21] The Rehabilitation Act, among other things, prohibits discrimination on the basis of disability by the federal government, federal contractors, and by recipients of federal financial assistance.[22] The ADA establishes certain

---

[21] Under the Rehabilitation Act and the ADA, an individual with a disability includes any person who has a physical or mental impairment that substantially limits one or more major life activities, has a record of such an impairment, or is regarded as having such an impairment.

[22] 29 U.S.C. § 701 *et seq.* Section 504 of the Rehabilitation Act provides that no otherwise qualified individual with a disability shall, solely on the basis of the disability, be excluded from the participation in, denied the benefits of, or be subjected to, discrimination under any program or activity that receives federal financial assistance, or any program or activity conducted by federal executive agencies. 29 U.S.C. § 794. Section 508 generally requires federal agencies to ensure that their electronic and information technology is accessible to individuals with disabilities, including employees and the public. In particular, it requires that members of the public with disabilities seeking information or services from a federal agency have access to and use of information and data that is comparable to the access and use by members of the public who do not have disabilities. 29 U.S.C. § 794d. In 2013, DHS issued Directive 065-01, Nondiscrimination for Individuals with Disabilities in DHS-Conducted Programs and Activities (Non-Employment), which included a requirement for DHS components – including FEMA – to (1) conduct a self-evaluation of their programs and activities to identify any barriers to access and any gaps in existing component policies or procedures for providing reasonable accommodations; and (2) develop a plan that addresses any identified barriers and documents the components' disability policies.

Disaster Assistance

161

non-discrimination and other requirements for employers, state and local governments, public accommodations, and telecommunication services with respect to people with disabilities.[23] In addition, the Post-Katrina Emergency Management Reform Act of 2006 (Post-Katrina Act) amended the Stafford Act to prohibit discrimination on the basis of disability by personnel carrying out federal assistance functions at the site of a major disaster, including distributing supplies, processing applications, and other relief and assistance activities.[24]

Inclusive practices are intended to ensure people with disabilities have equal opportunities to participate in, and receive the benefits of, emergency management programs and services. Such practices could include involving people with disabilities in emergency evacuation planning, ensuring that shelters are physically accessible, and providing guidance on post-evacuation residency for individuals with disabilities. FEMA's administrator appointed a Disability Coordinator to comply with a requirement in the Post-Katrina Act.[25] Under the Post-Katrina Act, the Disability Coordinator is responsible for ensuring coordination and dissemination of best practices and the development of training materials for emergency managers on the needs of individuals with disabilities, among other things. In 2010, FEMA created the Office of Disability Integration and Coordination (ODIC), which was designed to promote inclusive practices for assisting disaster survivors with disabilities.[26]

---

[23] 42 U.S.C. § 12101 *et seq.*

[24] See 42 U.S.C. § 5151(a). Other laws also establish protections for individuals with disabilities that may apply to federal, state, local, and/or private organizations involved in disaster response, such as the Fair Housing Act of 1968, as amended.

[25] See 6 U.S.C. § 321b.

[26] We reported in 2017 that despite steps FEMA had taken to improve disaster services for people with disabilities, it had not established procedures to help ensure regions consistently involve FEMA's Office of Disability Integration and Coordination (ODIC) in their disability integration activities. We also reported that ODIC's approach to delivering disability integration training was limited. We recommended that FEMA establish written procedures for involving ODIC in regional activities; set goals for disseminating disability integration training; and evaluate alternative delivery methods for the training. FEMA agreed with the recommendations, and has efforts underway to address them. See GAO, *Federal Disaster Assistance: FEMA's Progress in Aiding Individuals with Disabilities Could Be Further Enhanced,* GAO-17-200 (Washington, D.C.: Feb. 7, 2017).

162　　*United States Government Accountability Office*

Other laws establish non-discrimination protections for individuals based on age, which may also apply to federal, state, local, or private organizations involved in emergency management. Specifically, the Stafford Act prohibits discrimination on the basis of age by personnel carrying out federal assistance functions at the site of a major disaster.[27] In addition, the Age Discrimination Act of 1975, as amended, provides that no person shall, on the basis of age, be excluded from participation in, be denied the benefits of, or be subjected to discrimination under, any program or activity receiving federal financial assistance.[28]

Federal Assistance for Individuals with Disabilities, Including those Who Are Older, In Response to the 2017 Hurricanes

FEMA staff who are responsible for focusing on inclusive practices in emergency management include those deployed to areas affected by disasters and those working permanently in FEMA's regional offices. In 2017 and earlier, FEMA deployed staff called disability integration advisors (DIA) to identify and recommend courses of action to meet the needs of disaster survivors with disabilities. DIAs were expected to help individuals with disabilities, including those who are older, with accessing assistance such as temporary housing, medical equipment, food, and shelter. They were also expected to work beside other FEMA staff who worked directly with individuals, such as those on Disaster Survivor Assistance Teams, which go door-to-door to meet people in their communities.

In 2017 and earlier, disability integration staff working permanently in FEMA's regional offices, known as Regional Disability Integration Specialists (RDISs), were responsible for promoting inclusive practices through outreach to state and local emergency managers in all locations, whether or not they had recently been affected by a disaster. RDISs were also generally expected to track information about service and support

---

[27] See 42 U.S.C. § 5151(a).
[28] See 42 U.S.C. § 6102.

Disaster Assistance   163

shortfalls and demographics in each FEMA region, such as local and state statistics on individuals who are deaf or hard of hearing, to help keep track of what FEMA should be prepared to address in a disaster. According to FEMA officials, the specific job responsibilities of a RDIS, both prior and subsequent to 2017, may vary from region to region, as each Regional Administrator determines the staffing structure that meets the emergency management needs for his or her region.[29]

Revised Approach for Assisting Individuals with Disabilities, Including Those Who Are Older

In June 2018, FEMA officials began implementing a new approach to disability integration. According to FEMA officials, this new approach was intended to incorporate lessons learned from the 2017 hurricane season.

Disability Integration Staff in the Regions

FEMA officials recommended that Regional Administrators add new disability integration staff in each of the regions to foster day-to-day relationships with state and local emergency managers and disability partners (see Table 1). These new staff, whom FEMA also refers to as DIAs, would work with state and local emergency management agencies to advise and support emergency managers on inclusive practices. These regionally-placed DIAs would help develop relationships between state and local emergency managers and the disability community to build capacity and a culture of preparedness. Under the new disability integration approach, the role of the RDIS remains largely the same with the exception of coordinating with regionally-placed DIAs on disability integration.

---

[29] Regional Administrators in each of the regional offices report directly to the FEMA Administrator and are responsible for the day-to-day management and administration of regional activities and staff.

164　　　*United States Government Accountability Office*

Table 1. Former and New Models for Staffing FEMA's Disability Integration Approach in the Regions

Staff		Former Model (2017)	New Model (2018)
New position: Disability Integration Advisors (DIAs) assigned to work with state and local emergency managers	*Number and location*	Not Applicable	• DIAs assigned to work with state and local emergency managers in coordination with RDISs
	Role	Not Applicable	• Support state and local emergency managers by advising on inclusive emergency management principles and practices • After a disaster declaration, advise the Federal Coordinating Officer on capabilities of impacted area and leverage existing relationships with on-the-ground partners
Regional Disability Integration Specialists (RDISs)	*Number and location*	• One RDIS in each of the 10 FEMA regions	• One RDIS in each of the 10 FEMA regions
	Role	• Promote inclusive practices to local emergency managers • Track service and regional statistics on individuals with disabilities to inform disaster preparedness • Deploy to disaster locations • Conduct outreach, education, and training to state and local partners; support FEMA regional program offices	• In addition to their previous role, RDISs serve as team lead for DIAs working in state partnership

Source: GAO analysis of Federal Emergency Management Agency (FEMA) documentation and interviews with FEMA officials as of April 2019. | GAO-19-318.

Deployment of Disability Integration Staff

FEMA officials also made changes to how the agency deploys disability integration staff (see Table 2). The new deployment model for disability integration staff was designed to shift the responsibility of assisting individuals with disabilities from DIAs to all FEMA staff, such as those registering people for individual assistance. To implement this new model, FEMA plans to train all the agency's deployable staff and staff in programmatic offices on disability issues during response and recovery deployments. According to FEMA, a smaller number of DIAs would deploy to advise FEMA leadership in the field during disaster response and recovery.

Table 2. Former and New Models for Deploying FEMA Disability Integration Staff

Staff	Former Model (2017)		New Model (2018)
Deployed Disability Integration Advisors (DIAs)	*Number deployed per disaster*	• Average peak deployment of 55 DIAs to 2017 hurricane affected locations	• Start with 5, scale up as needed
	Role	• Assist federal, state, and local emergency managers on accessibility issues • Directly assist individuals with disabilities, including those who are older	• Advise Joint Field Office leadership and programmatic offices, including Individual Assistance Branch Chief, among others, on integrating disability principles throughout FEMA's response and recovery efforts • No longer providing direct assistance to individuals with disabilities
Other deployed FEMA staff, including from Disaster Survivor Assistance Teams	*Role*	• Partner with DIAs to assist disability-related cases	• After receiving training on disability competencies, directly assist individuals who are older or have disabilities

Source: GAO analysis of Federal Emergency Management Agency (FEMA) documentation and interviews with FEMA officials as of April 2019. | GAO-19-318.

As discussed later in this chapter, FEMA deployed disability integration staff under its new model for 2018 disasters, but implementation of FEMA's new model for disability integration in the regions was ongoing as of May 2019.

166   *United States Government Accountability Office*

OFFICIALS FROM STATES, TERRITORIES, LOCALITIES, AND NONPROFITS IN OUR REVIEW REPORTED CHALLENGES PROVIDING TARGETED, TIMELY ASSISTANCE TO INDIVIDUALS WHO A ARE OLDER OR HAVE DISABILITIES

State, Territorial, Local, and Nonprofit Officials Reported Difficulty Providing Critical Goods and Services

Territorial, local, and nonprofit officials from all four locations in our review, as well as survivors we interviewed from the U.S. Virgin Islands, reported that the substantial damage caused by the 2017 hurricanes prevented or slowed critical assistance for some individuals who are older or have disabilities. State, territorial, and local governments and nonprofit organizations provided some of this assistance—such as food, water, and medical care—to the general population following the hurricanes. However, some individuals who are older or have disabilities may have required specialized assistance, such as help filling medication prescriptions or equipment to maintain health and independence. Many of the challenges associated with responding to individuals' basic needs were exacerbated in Puerto Rico and the U.S. Virgin Islands. As we previously reported, a number of major factors affected response to Hurricanes Irma and Maria—including debilitated power grids, communication systems, and transportation infrastructure—and limited local preparedness for large, consecutive hurricanes.[30]

Food and Water

According to territorial and nonprofit officials in Puerto Rico and the U.S. Virgin Islands, as well as survivors we interviewed in the U.S. Virgin Islands, some individuals who are older or have disabilities experienced difficulty obtaining needed food and water due to centralized distribution

---

[30] GAO-18-472.

Disaster Assistance 167

models. For example, Maria significantly diminished transportation capabilities in Puerto Rico and, according to officials from one nonprofit organization, the majority of food and water was distributed to centralized locations around the island as a result. Those officials said this posed a major barrier to people with mobility challenges or without caregivers receiving food and water. Officials from one governmental agency in Puerto Rico reported that some had to rely instead on local or federal officials to deliver these items to their homes, which took time and in some cases, did not happen. In addition, nonprofit officials and survivors we interviewed in the U.S. Virgin Islands indicated that the food available posed medical dietary risks to some older people and people with disabilities because it was high in sugar, fat, and salt.[31]

Medication and Medical Care

Government and nonprofit officials in Puerto Rico and the U.S. Virgin Islands, as well as individual survivors themselves, reported that individuals who are older or those with disabilities sometimes had difficulty accessing resources critical to sustaining health. In particular, territorial and nonprofit officials and individuals we spoke to highlighted challenges Maria survivors faced in obtaining needed medication—as well as oxygen—in Puerto Rico and the U.S. Virgin Islands. Territorial health officials in Puerto Rico explained that the sustained power outage and diminished transportation capabilities affected Puerto Rico's two oxygen producers' ability to sufficiently supply the island.[32] Also, according to officials from one Puerto Rico nonprofit, individuals who needed electricity to power respirators or continuous positive airway pressure machines had to rely on gas-powered generators which were also in short

---

[31] According to the Academy of Nutrition and Dietetics, diets high in sodium, high-fat dairy products, and sweets can increase the risk of hypertension and mortality. Older adults, who often have multiple medical conditions requiring them to alter their dietary intake, can effectively manage these conditions through nutrition.

[32] In its after-action report/improvement plan, the Puerto Rico Department of Health reported that it did not have enough assets to fulfill healthcare facilities' requests for medical supplies, including oxygen. Government of Puerto Rico, Department of Health, *Hurricanes Irma and Maria, PRDH Emergency Operations, After-Action Report/Improvement Plan*, Rev. 2017 508 (Dec. 20, 2017).

168 *United States Government Accountability Office*

supply. In addition, one Independent Living Center participant we interviewed in the U.S. Virgin Islands told us that while he was able to get one of his regular prescriptions from a local clinic, none of the pharmacies on the island had the stock to fill it. A second participant told us that the price of the medication she needed to treat her diabetes had more than doubled since Maria, which she attributed to a lack of supply.

According to territorial officials from the U.S. Virgin Islands and officials from one Puerto Rico nonprofit, a number of local medical facilities— including hospitals—in the U.S. Virgin Islands and Puerto Rico sustained substantial damage, impairing their ability to treat the local community. Officials from one nonprofit in the U.S. Virgin Islands said that some individuals were admitted to hospitals and nursing homes off the island despite needing only routine treatment, such as dialysis. In Puerto Rico, hospitals lacked clinical staff and treated only those with the most critical health concerns, according to officials from one nonprofit, leaving some individuals without treatment.

Shelter Services, Disaster-Related Information, and Transportation

Individuals who are older or have disabilities affected by the 2017 hurricanes also may have faced challenges accessing services from local shelters, according to state, territorial, local, and nonprofit officials we interviewed in all four locations. Officials from each of the four disaster locations reported examples of challenges accessing basic shelter services, including restrooms and food. For example, nonprofit officials in Florida and Puerto Rico described instances of shelter residents holding up sheets around other residents whose impairments prevented them from accessing the restrooms so that they could relieve themselves in common spaces of the shelter.

Disaster-related information, which equips those affected by a disaster to make decisions about evacuating, sheltering, and returning home, was inaccessible to people with certain disabilities, according to local and nonprofit officials we interviewed in all four disaster locations.[33] In

---

[33] Throughout this report, we use the terms "accessible" or "inaccessible" as they were used in our interviews or in documents we reviewed; however, we did not independently evaluate

interviews with these officials we heard multiple examples of the challenges faced by people with hearing impairments to getting timely and accurate information. For example, nonprofit officials in the U.S. Virgin Islands explained that the only source of disaster-related information there was the radio, so residents with hearing impairments had to rely on friends and neighbors to relay the information.[34]

This sidewalk connected a busy bus stop to the island's only hospital.
Source: GAO. | GAO-19-318.

Figure 1. Storm Damage in St. Croix, the U.S. Virgin Islands.

State, territorial, local, and nonprofit officials we spoke to reported that transportation was a substantial challenge for all of the affected locations following the 2017 disasters, but was especially challenging for those who relied on public transportation or were unable to walk long distances, such as people who were older or have disabilities (see Figure 1). Florida state officials reported that few public transportation services, including

---

whether any programs or activities discussed in this report met any applicable legal standards for accessibility.

[34] According to one FEMA official, FEMA made efforts to reach Virgin Islanders with hearing impairments. For example, FEMA made interpreters available at Disaster Recovery Centers, at FEMA press conferences, and on social media.

paratransit, were functional following Hurricane Irma. This may have prevented some people who are older or have disabilities from evacuating to shelters before Irma and from maintaining their health and wellness—for example, by shopping for groceries or going to medical appointments—after the storm, according to state officials.

Officials from Puerto Rico and the U.S. Virgin Islands Said They Faced Challenges Locating People Who are Older or Have Disabilities

Officials we spoke to from territorial governments and nonprofits said they had difficulty locating people who are older or have disabilities following the 2017 hurricanes, which made providing disaster assistance challenging. In Puerto Rico, nonprofit and territorial government officials involved in responding after Irma and Maria explained that at that time no government entity maintained a list of nursing homes for people who are older or have disabilities. Because communication capabilities were nonexistent for a number of weeks across most of the island, these officials explained that government and nonprofit staff were dispatched as a task force to physically locate these facilities. Task force members identified individuals living at home or in residential facilities, conducted wellness checks, and distributed food, water, and other goods. According to territorial and nonprofit officials, a very small percentage of older Puerto Ricans live in nursing homes, while most live in their own homes, which can be difficult to reach. Officials in Puerto Rico reported identifying 503 licensed nursing homes, with an average of approximately 20 residents per facility, according to a list one territorial government agency compiled in early 2018. One nonprofit official noted that the search for those in institutions took more than a month.

Territorial and nonprofit officials in the U.S. Virgin Islands reported similar difficulties, with one territorial official explaining that residents with disabilities who do not access government services are a challenge to identify, and therefore, assist. One nonprofit official described

Disaster Assistance 171

commissioning a bus to travel around the island of St. Croix to identify people with disabilities in need, most of whom lived alone. According to one territorial official, the challenge of identifying people was compounded by a general stigma associated with disabilities in the U.S. Virgin Islands. In addition, some Virgin Islanders with disabilities protected under the ADA may not self-identify as a person with a disability, according to FEMA officials we spoke with who were deployed there. To address these challenges, territorial and nonprofit officials in Puerto Rico and the U.S. Virgin Islands, as well as an individual survivor in the U.S. Virgin Islands, told us they supported the idea of some type of registry responders could use to locate people who are older or have disabilities.

Texas and Florida operate registries to help local governments prepare to assist residents with disabilities during evacuations and sheltering, but state, local, and nonprofit officials in those locations reported that confusion about the registries limited their effectiveness. According to state officials in Texas, that state's registry provides local emergency management officials with information about the needs in their community, especially with regard to individuals with disabilities and other access and functional needs. However, representatives of state, local, and nonprofit agencies reported a general misconception among residents that signing up for Texas' registry would guarantee direct evacuation or transportation assistance. Florida's registry similarly provides local emergency managers with information to prepare for disasters, especially related to sheltering for people with disabilities. State residents who are registered may access special needs shelters, which are required by state law and typically provide electricity and oxygen, according to state, local, and nonprofit officials. However, state and nonprofit officials said residents were sometimes unsure about whether the special needs shelters were appropriate for them, which may have resulted in delayed registrations, residents registering who did not need to, or under-registration. Officials from Texas expressed concerns that confusion about the registries would lead to residents' overreliance on public disaster services. In addition, state and local officials in both states said a late

172　　　*United States Government Accountability Office*

influx of registrations complicated their states' response efforts. Texas state officials said they saw a 200-percent spike in registrations in the days after Hurricane Harvey was forecast to hit the state, while officials from one Florida county reported that 800 county residents registered for the first time during the 4 days before Irma's landfall. Employees in that county tried to call each registrant directly to get more information about their needs, and the late registrations made this difficult, according to officials.

A Lack of Timely Data from FEMA Hampered Some Partners' Efforts to Target Assistance

Officials we interviewed from Texas, Florida, and Puerto Rico reported difficulty obtaining FEMA data that could help them deliver assistance to individuals, including those who are older or have disabilities.[35] These officials explained that data—including names and addresses—showing who has registered for and received Individual Assistance from FEMA's disaster assistance programs can help local governments and nonprofits identify who in their community needs assistance. For example, nonprofit officials in Puerto Rico said they could have more effectively provided people with disabilities donated goods sent from all over the world if they knew who requested similar items from FEMA. In addition, FEMA data can help jurisdictions identify which community residents with disabilities have not applied for FEMA assistance, and local officials said they can use this information to target individuals who may need help with FEMA's registration process. Finally, officials explained that FEMA's data could inform recovery efforts by helping local officials identify parts of the community that were

---

[35] Officials from the U.S. Virgin Islands did not note this as a challenge to their efforts to provide disaster assistance to individuals who are older or have disabilities.

Disaster Assistance 173

underserved by FEMA assistance or that may need assistance after a FEMA program ends.[36]

Texas officials from two counties said they made multiple requests for FEMA's registrant data, which they hoped to use to canvass areas of their counties to offer assistance, including helping people apply for FEMA financial assistance. One county eventually obtained data from FEMA more than 4 months following Harvey after applying constant pressure on FEMA, according to officials. Florida state officials reported that their request for information on older residents receiving FEMA Transitional Sheltering Assistance was denied. They intended to share it with affected counties that needed to plan sustainable housing options for those recipients when FEMA's program ended. According to officials, the denied request delayed counties' ability to plan for sustainable housing after the storm.

According to FEMA officials, the agency has broad authority to share its data on registrations, and follows the framework established under the Privacy Act of 1974 on the collection, use, maintenance, and dissemination of personally identifiable information.[37] Generally, FEMA uses two types of agreements—FEMA-State Agreements and Information Sharing Access Agreements—to establish the terms and conditions of how it will share its

---

[36] FEMA assistance may be subject to financial and/or time limits, depending on the program. For example, Transitional Sheltering Assistance is approved for an initial period of 5-14 days, which may be extended for up to 6 months. The Disaster Recovery Reform Act of 2018, among other things, increased the amount of assistance available to individuals with disabilities under IHP, including allowing accessibility repairs for people with disabilities without counting those repairs against their maximum disaster assistance grant award.

[37] The Privacy Act of 1974, as amended, generally requires federal agencies to obtain individuals' written consent before disclosing identifiable information about those individuals from a "system of records" maintained by the agency. See 5 U.S.C. § 552a(b). Under the act's "routine use" exception, however, an agency is permitted to disclose such records when used for a purpose compatible with the purpose for which they were collected. Each agency that maintains a system of records is required to publish a notice, known as the System of Records Notice, in the *Federal Register*, describing the information the agency collects and each routine use, including the categories of users and the purposes of each use. FEMA's most recent notice for its disaster assistance system of records lists a number of routine uses under which FEMA may disclose personally identifiable information to state, tribal, and local government agencies and emergency managers, including the type of information it can share and under what circumstances. See Privacy Act of 1974; Department of Homeland Security Federal Emergency Management Agency – 008 Disaster Recovery Assistance Files System of Records, 78 Fed. Reg. 25,282 (Apr. 30, 2013).

174	*United States Government Accountability Office*

data.[38] According to officials, these agreements are disaster-specific and are established after a disaster is declared. The agreements include a list of state departments, agencies, or specific individuals, known as "Trusted Partners," who are authorized to use FEMA data on behalf of the state or organization. According to state and nonprofit officials, however, obtaining FEMA data has sometimes been challenging and time consuming. According to one nonprofit official, for example, FEMA sent data 6 weeks after the hurricane made landfall. State officials we interviewed in Florida said the process of requesting and receiving data was slow, and that the delayed access to the data limited state officials' ability to assist individuals in a timely manner.

FEMA officials we interviewed said that the challenges nonfederal partners faced obtaining FEMA data on Individual Assistance registrations are likely due to insufficient or unclear communication between FEMA and the states. For example, they said that joint field office leadership, who are responsible for developing the FEMA-State Agreements, may not have sufficient time to work with state and local officials to compile a comprehensive list of Trusted Partners. They explained that FEMA-State Agreements are developed quickly after a disaster declaration, when multiple priorities compete on short timelines and FEMA officials may not be aware of all the state- and local-level entities and individuals who need FEMA data to provide assistance. One FEMA official described situations in which state contractors requesting data should have had access to it but, according to the official, because they were not named in the agreement,

---

[38] A FEMA-State Agreement sets forth the understandings, commitments, and conditions for FEMA assistance in a state or territory and is generally a prerequisite to all forms of FEMA assistance. See 44 C.F.R. § 206.44. Under FEMA's System of Records Notice, state agencies may request and receive information using the protocols established in an appropriate FEMA-state agreement, and they may also share information they receive from FEMA with their contractors/grantees, and/or agents that are administering a disaster related program on behalf of the agency under the same protocols. According to a FEMA policy document on data sharing, an Information Sharing Access Agreement is a "covenant between FEMA and non-FEMA parties that provides the terms and conditions of any information sharing." See FEMA Recovery Policy 9420.1, *Secure Data Sharing* (Sept. 9, 2013). For example, FEMA officials said they regularly use Information Sharing Access Agreements to share data with the American Red Cross. According to FEMA policy, if the stated purpose for the requested FEMA data is inconsistent with those outlined in FEMA's System of Records Notice, FEMA will reject the request.

Disaster Assistance 175

FEMA would not allow sharing it. Further, FEMA officials suggested that while states may modify their FEMA-State Agreements to add to the list of Trusted Partners, state officials may not be aware of this option and, according to the officials, do so infrequently.

In addition, officials said that states may request FEMA data through an Information Sharing Access Agreement in lieu of modifying their FEMA State Agreement, but FEMA does not provide guidance to partners on drafting or submitting these agreements. FEMA officials explained that FEMA program staff—rather than state officials—draft these agreements, so they did not see the need to issue guidance. However, officials added that these requests sometimes do not specify all the data elements FEMA needs, so staff has to work with the partners to refine the requests. Emergency management officials we interviewed in Florida reported that although nonprofit partners not named in the FEMA-State Agreement were denied their request for FEMA data, neither the state nor the nonprofit organizations requested data through Information Sharing Access Agreements, and the state did not request a modification of the FEMA-State Agreement.

FEMA officials described how ongoing agency efforts to improve overall processes could help address some of these data-sharing challenges. For example, officials reported that FEMA headquarters staff is developing tools to share with FEMA regional staff to help them standardize language in data-sharing agreements, which they said could expedite data sharing. They also told us that FEMA plans to issue a data management directive to FEMA staff to help ensure that internal and external stakeholders have the data they need and that the data are usable.[39]

Despite these efforts, FEMA has not made a concerted effort to educate nonfederal partners on FEMA's data sharing process, which could better facilitate Trusted Partners obtaining Individual Assistance data soon after disasters occur. FEMA officials said they could provide better instruction on the data sharing process on the agency website to help partners effectively request and work with FEMA to draft agreements, for

---

[39] As of February 2019, FEMA officials said that the directive was expected to be released in March 2019.

176 *United States Government Accountability Office*

example, by making available examples of approved agreements. Federal standards for internal control state that agency management should externally communicate the necessary information to achieve the entity's objectives. Without guidance from FEMA on how to effectively navigate its data-sharing process, nonfederal partners risk not receiving timely data. As a result, they may be hindered in their ability to contact and assist specific individuals, including people who are older or have disabilities, who need transportation, housing, disaster recovery, and other key types of assistance.

ASPECTS OF FEMA'S APPLICATION PROCESS FOR ASSISTANCE CREATED CHALLENGES FOR INDIVIDUALS WHO ARE OLDER OR HAVE DISABILITIES

Some Individuals Who Are Older or Have Disabilities Experienced Long Wait Times and Other Challenges Registering for FEMA Assistance

Individuals who are older or have disabilities faced challenges registering for assistance under FEMA's IA program over the phone, online, and in-person following the 2017 hurricanes, based on our analysis of FEMA data, review of relevant reports, and interviews with FEMA officials and stakeholders.[40]

FEMA Disaster Assistance Helpline

FEMA data indicate that individuals confronted long wait times when trying to apply for assistance through the agency's helpline, which may

---

[40] As previously mentioned, individuals applying for IA register by answering a set of standard questions that help determine their eligibility. After answering these registration-intake questions, registrants typically must take additional steps to complete the application process for FEMA assistance. For example, FEMA typically conducts a home inspection to assess damaged property and may request that registrants submit additional documentation to verify eligibility, such as proof of ownership and insurance information.

Disaster Assistance

pose greater challenges for those who are older or have a disability.[41] FEMA has a performance goal for its call centers to answer helpline calls within 20 seconds. However, FEMA officials acknowledged this goal may be unattainable during times of high call volume. In the days after Hurricane Maria affected the territories—and survivors from Harvey and Irma were concurrently contacting the helpline—up to 69 percent of calls went unanswered and the daily average wait time for answered calls peaked at almost an hour and a half, according to our analysis of FEMA data (see Figure 2).[42] The daily average wait time decreased following this peak and consistently remained under 30 minutes post-October 6, about 2 weeks after Hurricane Maria.

While long wait times could be burdensome for all individuals, state officials and disability advocates we interviewed said long wait times were especially burdensome for people with certain disabilities, such as those with attention disorders or whose assistive technology prevents multi-tasking when waiting on hold. Individuals who rely on paid caregivers or interpreters to communicate also may have incurred additional costs as a result of the extra assistance they required while waiting on hold, according to a disability advocate. In addition, a Florida state official who works with older adults explained that many of these individuals may be on a fixed income and could not afford to wait on hold because their cell phone plans had limited minutes available each month.

FEMA officials told us that the number of calls to the helpline after the 2017 hurricanes was overwhelming and created an unprecedented need for additional staff to support the high call volume.

---

[41] Individuals can call FEMA's helpline to complete their registration for FEMA assistance, ask questions about the application process, or check on the status of an existing application.

[42] FEMA calculates the daily average of wait times for all answered calls. As a result, some callers may experience either shorter or longer wait times than the average. State officials, a disability rights organization's report, and a participant at a listening session hosted by CRCL and FEMA reported that some individuals waited on hold for over 6 hours before their call was answered by a FEMA helpline representative. According to FEMA, long wait times are associated with higher rates of unanswered calls because individuals either hang-up or the call is involuntarily disconnected—such as from a cell phone running out of batteries. FEMA does not track wait times for callers who hang up or are otherwise disconnected before reaching a call center representative.

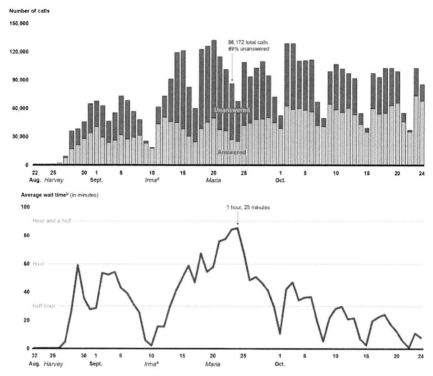

Source: GAO analysis of Federal Emergency Management Agency (FEMA) data. | GAO-19-318.

[a]As a result of Hurricane Irma, major disasters were declared in the U.S Virgin Islands on September 7, 2017 and in Puerto Rico and Florida on September 10, 2017.

[b]Wait time represents the average time period before calls were answered each day. The Federal Emergency Management Agency (FEMA) does not track wait times for callers who hang up or are otherwise disconnected before reaching a call center representative. FEMA's count of unanswered and total calls excludes calls that were disconnected before 20 seconds elapsed. FEMA considers these voluntary disconnections that are not due to extended wait times.

Figure 2. Volume of Calls and Average Daily Wait Times for the Federal Emergency Management Agency (FEMA) Disaster Assistance Helpline, August 22-October 24, 2017.

As a result, FEMA officials told us they sought to reduce wait times by adding two fully staffed external call centers in September 2017 and maintaining the increased staffing to support the helpline during the 2018 hurricane season.

Online and Mobile Application

Most individuals who are older or reported having disabilities ultimately registered online or through the mobile application (app); however, local officials and others reported that some of these individuals had difficulty using these methods.[43] Some individuals who are older or have disabilities do not have smart phones or internet access and others may not be computer literate or proficient in technology, according to FEMA, local, and nonprofit officials and participants of a listening session in Florida organized by the DHS Office for Civil Rights and Civil Liberties (CRCL) and FEMA. Further, the online application requests that registrants provide an email address, but some older adults do not have one or do not use one often, according to local officials. As a result, these individuals may not receive critical correspondence from FEMA. Also, FEMA's 2017 internal self-evaluation found the mobile app, which provides individuals with registration information, among other types of disaster-related information, was not compliant with Section 508 of the Rehabilitation Act.[44] In 2018, FEMA conducted a compliance test that found features of the mobile app remained inaccessible for people with blindness and low vision, according to FEMA officials.[45] Additionally, prolonged loss of power and lack of internet access and phone connectivity compounded challenges to registering with FEMA online and over the phone for all individuals, including those who are older or reported having disabilities. For example, in Puerto Rico, widespread damage from

---

[43] Specifically, 84 percent of registrants who reported a disability to FEMA and 83 percent of registrants age 65 or over applied online or through the mobile application. Comparatively, 92 percent of registrants under age 65 applied online or through the mobile application.

[44] As previously discussed, section 508 of the Rehabilitation Act generally requires federal agencies to ensure that their electronic and information technology is accessible to individuals with disabilities, including employees and the public. 29 U.S.C. § 794d. FEMA completed its self-evaluation report in August 2017 after a six-month assessment to evaluate its facilities, programs, policies, and practices and determine how effectively the agency provides equal physical, program, and effective communication access to people with disabilities.

[45] FEMA officials said that in 2019 the agency plans to remediate the identified compliance issues and test improvements for the mobile application. Officials said a user experience group, including those who have access and functional needs or use assistive technologies, will inform the mobile application's design and development.

180 *United States Government Accountability Office*

hurricane Maria left 3.7 million residents without electricity and 95 percent of cell towers out of service for a prolonged period of time.[46]

Disaster Recovery Centers

Lack of transportation and long lines, among other issues, affected some individuals' ability to apply for FEMA assistance in person at Disaster Recovery Centers, according to FEMA officials, a disability rights organization report, and others. Some people with disabilities, particularly those in Puerto Rico and the U.S. Virgin Islands, reportedly could not travel to Disaster Recovery centers because they lacked conventional or accessible transportation or the roads were inaccessible, according to a disability rights organization's report.[47] Further, a participant in a CRCL and FEMA listening session in Florida said the local Disaster Recovery Center was in an isolated location far from public transportation. Other individuals with disabilities who rely on power to operate life-sustaining equipment could not leave their homes to travel to the centers, according to the disability rights organization's report.

FEMA, state, and nonprofit officials also stated that individuals in some locations were required to wait in long lines to register for FEMA assistance, which was especially challenging for individuals who are older or have certain disabilities, such as those whose disability impacts their ability to stand. According to the disability rights organization's report, Disaster Recovery Centers in some locations did not have qualified on-site or remote sign language interpreters and did not always have print or electronic information in accessible formats for people with vision

---

[46] According to local utilities, it took roughly 5 months for power to be restored to all the customers able to receive power safely in the U.S. Virgin Islands, and roughly 11 months for power to be restored to all customers able to receive power safely in Puerto Rico. The electric utility in the U.S. Virgin Islands serves approximately 55,000 customers and the electric utility in Puerto Rico serves approximately 1.5 million customers. Long-term infrastructure outages in Texas and Florida were isolated. According to FEMA, 10 days after Hurricane Harvey's landfall, 55,000 customers in Texas were without power, reduced from a peak of approximately 300,000 customers. In Florida, 75,000 customers were without power 10 days after Hurricane Irma made landfall, reduced from a peak of more than six million customers.

[47] The Partnership for Inclusive Disaster Strategies. *Getting It Wrong: An Indictment with a Blueprint for Getting It Right. Disability Rights, Obligations and Responsibilities Before, During and After Disasters* (May 2018).

Disaster Assistance 181

disabilities or low literacy. According to FEMA officials, some buildings— particularly in the U.S. Virgin Islands and Puerto Rico—used by FEMA to provide disaster assistance, including Disaster Recovery Centers, required substantial FEMA efforts to make them accessible. For example, they said that in the U.S. Virgin Islands FEMA had to make significant changes to improve accessibility at Disaster Recovery Centers, such as shipping ADA compliant toilets there, paving the parking lot to add accessible parking spots, and building an accessible ramp to the entrance of the building. Consistently opening accessible facilities remains an area for improvement for the agency, according to FEMA's internal self-evaluation. FEMA's mission requires swift action once a disaster occurs and the lack of available accessible facilities and competing priorities kept FEMA from consistently opening accessible facilities, according to the evaluation.

Disaster Survivor Assistance Teams

Disaster Survivor Assistance Teams helped some individuals who are older or have disabilities access FEMA's registration process. According to FEMA officials, these teams primarily address the needs of people disproportionately affected by disasters, including individuals who are older or have disabilities. One of their tasks is helping individuals, including those who cannot otherwise register over the phone, online, or at a Disaster Recovery Center, register for FEMA assistance.[48] FEMA data show that 2 percent of all older adults who registered for FEMA assistance after the 2017 hurricanes did so with help from the Disaster Survivors Assistance Teams, compared to 0.7 percent of applicants under the age of 65.[49] Some FEMA staff in the U.S. Virgin Islands consistently went into the field to locate and assist people who could not leave their homes, according to the disability rights organization's report. However,

---

[48] Disaster Survivor Assistance Teams are equipped with tablets and other mobile tools that allow them to register individuals at their home, shelter, or wherever they may be, according to FEMA.

[49] Similarly, 1.4 percent of all registrants with a reported disability who registered for FEMA assistance after the 2017 hurricanes did so with help from the Disaster Survivors Assistance Teams, compared to 0.9 percent of registrants with no reported disability.

182   *United States Government Accountability Office*

communication issues in Puerto Rico limited the teams' effectiveness. According to FEMA, limited cellular service required teams to use paper forms or offline tablets and laptops to register individuals, which they reported caused inaccuracies and omissions that may have delayed benefits. Further, assisting individuals this way is more resource-intensive for FEMA than other registration options.[50] Despite these challenges, Disaster Survivor Assistance Teams were helpful in registering individuals, particularly when they were accompanied by officials from local disability organizations, FEMA DIAs, and qualified sign language interpreters, according to the disability rights organization's report.

FEMA Did Not Provide Individuals Clear Opportunities to Disclose Disability-Related Needs and Request Accommodations during the Application Process

FEMA's registration process does not give individuals a clear opportunity to state they have a disability or request an accommodation. FEMA primarily collects information about individuals' disability-related needs from standard questions asked during the registration process.[51] FEMA also provides accommodations to individuals with a disability or

---

[50] We previously reported that in the aftermath of Hurricane Maria, officials needed to conduct more door-to-door visits to reach disaster survivors and conduct assessments compared to the continental United States where individuals mostly apply for IA at Disaster Recovery Centers or online, according to FEMA officials. Locating addresses and individuals in Puerto Rico was challenging, according to FEMA officials, because many affected areas did not have posted addresses, many individuals use nicknames instead of their given names, and often several families were located on a single property. In addition, FEMA did not have enough bilingual employees to communicate with local residents or translate documents. According to FEMA officials, this resulted in further delays while staff were reshuffled from other disasters to Puerto Rico. GAO-18-472.

[51] For the purposes of this report, we use the term "disability-related needs" broadly to include all needs individuals may have that are related to a disability or access or functional need. For example, this may include replacement of a damaged wheelchair or other durable medical equipment, fixing an accessible ramp to a house, or any needed assistance to perform daily activities—such as showering, getting dressed, walking, and eating. Individuals with such needs may be eligible for FEMA assistance; for example, IHP provides financial assistance for disaster-related medical expenses, including repair or replacement of medical equipment. FEMA may also refer people with disability-related needs to other agencies or non-profits for assistance.

Disaster Assistance

access and functional need to ensure they can complete the registration and application process and participate fully in FEMA programs.[52] However, FEMA's registration does not include questions that directly ask registrants if they have a disability or if they would like to request an accommodation for completing FEMA's application process. See Figure 3 for the sequence of questions FEMA uses to identify if someone needs assistance as a result of a disability.

FEMA officials stated that the disability-related questions are intended to determine if an individual has a disability and to identify disability-related needs, such as an accommodation to participate in the registration process. The information is intended to help FEMA staff match individuals with disabilities with appropriate resources in a timely and efficient manner and target additional assistance, such as individualized calls to help with the application process. According to FEMA, the questions are an acknowledgement that disability-related needs during a disaster can have life-saving implications and require additional attention. The information also informs policy and strategy for providing assistance to people with disabilities, according to a FEMA official.

However, the disability-related questions can be difficult to interpret, according to state and local officials and disability advocates. FEMA officials we interviewed and FEMA's internal self-evaluation also acknowledged the questions are unclear, consistently misinterpreted, and do not solicit accommodation requests or effectively collect information on an individual's disability and related needs.

---

[52] According to a FEMA policy document, FEMA makes reasonable accommodations to policies, practices, and procedures to ensure physical, programmatic, and effective communication access to FEMA disaster assistance. This may include using technologies and services to ensure effective communication with applicants with disabilities and others with access and functional needs. For example, FEMA provides sign language interpreters and materials in alternate formats (such as Braille, large print and electronic formats) upon request. FEMA also has amplified telephones, phones that display text, and amplified listening devices for people with hearing loss. Some individuals with access and functional needs may be entitled to non-discrimination and other protections, including reasonable accommodations, under applicable civil rights laws. However, in this report we do not assess whether any accommodations or other services provided by FEMA or its partners during the period covered by our review complied with any such laws.

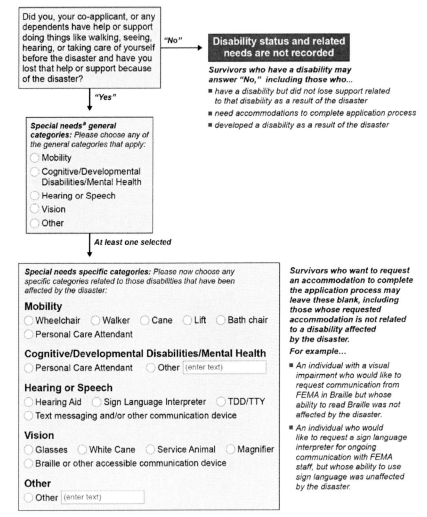

Source: GAO analysis of Federal Emergency Management Agency (FEMA) documentation. | GAO-19-318.

[a]The term "special needs" no longer aligns with FEMA's language standard for respecting the personal dignity of people with disabilities by using appropriate terminology, according to FEMA's internal self-evaluation.

Figure 3. Sequence of Disability-Related Questions in the Federal Emergency Management Agency (FEMA) Registration Process.

Disaster Assistance

For example, while the questions ask registrants whether they have lost "help or support," FEMA's internal self-evaluation stated that the broadness of that phrase confuses registrants. Such language could be unclear for individuals with disabilities who live independently, as well as for individuals with disabilities that lost durable medical equipment but not access to a caregiver, according to FEMA. Additionally, individuals with disabilities who have not lost the help and support they need may answer "no." These registrants will not be counted or identified as people with disabilities, even though they may still require additional assistance from FEMA, such as a sign-language interpreter or other accommodations to complete the application process. Moreover, answering "no" also prevents registrants from being asked additional questions that collect more detailed information about the nature of their disability or disabilities. Even if registrants answer "yes," the additional registration questions ask specifically about needs related to disabilities that have been affected by the disaster, which may differ from disability-related needs they may have to complete the application process. For example, an individual with a visual impairment who needs communication from FEMA in Braille or large print may not check any of the categories under "vision" if their ability to read Braille or large print was not directly affected by the disaster.

As a result of the unclear questions, individuals with disabilities may not request accommodations or report having a disability and related needs during FEMA's registration-intake. FEMA's internal self-evaluation reported that the wording of the disability-related questions is confusing and registrants most often leave them unanswered or provide inaccurate information, resulting in delays in the assistance process. The self-evaluation also found that if a registrant indicates a disability-related need by answering "yes," the registration process does not collect enough information to determine the nature of the need, such as whether it relates to a reasonable accommodation request or disaster assistance claim. As a result, the self-evaluation found that needs are not consistently matched to the appropriate FEMA disaster assistance program or service.

186   *United States Government Accountability Office*

FEMA provided additional services to registrants who responded to the disability-related questions following the 2017 hurricanes; however, its ability to target these services may have been negatively affected due to the unclear disability-related questions. Further, the internal FEMA self-evaluation concluded that the disability-related registration-intake questions do not yield consistent and accurate information about the needs of individuals with disabilities. Federal internal control standards call for agencies to use accurate information to achieve their objectives.[53] The information collected in the registration process may under-identify people with disabilities. For example, Puerto Rico has the highest estimated percentage of people with disabilities compared to any state or territory, 21.6 percent, according to 2017 Census data. However, less than 3 percent of all registrants in the territory answered "yes" to the disability-related question in response to hurricanes Irma and Maria.[54]

Other locations affected by the 2017 hurricanes had a similar pattern.[55] According to FEMA officials, while not all registrants with a disability

---

[53] GAO-14-704G.

[54] This analysis does not assess compliance with any applicable non-discrimination or civil rights laws. The data are from the 2017 Puerto Rico Community Survey, a survey administered annually by the United States Census Bureau. The Puerto Rico Community Survey produces 1-year estimates for the total civilian noninstitutionalized population and is the equivalent of the American Community Survey for the 50 states and District of Columbia. Data results from both surveys are released together as a unified American Community Survey dataset. The estimate for Puerto Rico has a margin of error at the 90 percent confidence interval of plus or minus 0.5 percentage points.

[55] Specifically, the estimated percentage of people with disabilities in Florida and Texas was 13.6 percent and 11.4 percent, respectively. However, 2.0 percent of registrants in Florida and 3.2 percent of registrants in Texas answered "yes" to the disability-related question. In the U.S. Virgin Islands, the estimated percentage of people with disabilities was 15.3 percent. However, 3.5 percent of all registrants in the territory answered "yes" to the disability-related question when applying for assistance after hurricanes Irma and Maria. The estimated percentage of people with disabilities for Texas and Florida was obtained from the 2017 American Community Survey, administered by the United States Census Bureau. Estimates from this source has margin of error at the 90 percent confidence intervals of plus or minus 0.5 percentage points or less. Data from the U.S. Virgin Islands was obtained from the 2010 decennial Census. All surveys asked about six disability types: hearing difficulty, vision difficulty, cognitive difficulty, ambulatory difficulty, self-care difficulty, and independent living difficulty. Respondents who report any one of the six disability types are considered to have a disability. Percentages reported by Census reflect the total civilian noninstitutionalized population for the entire state or territory. Percentages calculated using FEMA data reflect those who registered for FEMA's IA program. This analysis does not assess compliance with any applicable nondiscrimination or civil rights laws. Although not all who identify as having a disability for the Census survey have disability-related needs after disasters, the large

Disaster Assistance 187

have a related need, the large difference between Census disability data and FEMA's identification of registrants with disability-related needs illustrates the extent of potential under-counting.[56]

To address this potential under-counting, FEMA's disability integration staff spends a substantial percentage of their time trying to identify individuals with disabilities who may need additional assistance, according to a disability integration advisor. For example, FEMA staff created a list of approximately 100 key words that may indicate a disability and routinely search the notes field of case files to identify registrants that answered "no" or did not answer the disability-related questions but may have indicated the potential need for disability-related assistance.[57] In Puerto Rico, they identified an additional 94,000 cases of individuals that may have a disability or related needs using this process, but said more individuals likely could have benefited from additional assistance from FEMA, according to FEMA officials. Disability integration officials also emphasized the importance of interacting with individuals in their communities to address challenges they had reporting disability-related needs during the registration process. These interactions resulted in FEMA identifying additional individuals with unreported disabilities and related needs. According to a local official, some older individuals may not have self-identified as having a disability even if they needed assistance, such as for low vision.

In addition to the registration questions and outreach efforts, such as searching registrants' case files, FEMA may identify disability-related needs in other ways, such as through conversations with survivors during

---

difference between the two data sets illustrates potential under-counting in FEMA's data. In Texas and Florida, a portion of the state was not declared eligible for IA, meaning that the percentage of individuals with disabilities in the affected areas may differ from the percentage of individuals with disabilities in the entire state.

[56] FEMA's internal self-evaluation found similar disparities for a 2016 disaster when comparing the percentage of registrants that answered "yes" to the initial disability-related question to Census data on individuals reporting a disability in the disaster affected area. The evaluation concluded that while not every registrant with a disability is expected to answer "yes" to the disability-related question, the difference between the disability-related information collected during registration-intake and the Census is significant and should provide a basis for further investigation.

[57] In FEMA's April 2019 comments on this draft report, one official stated that this procedure is not a standard practice and has been discouraged by ODIC leadership.

188   *United States Government Accountability Office*

the registration process and when individuals actively request accommodations. FEMA officials stated that while registration-intake specialists are partially responsible for identifying individuals' disabilities, registrants are also responsible for disclosing their disability-related needs. However, some individuals may not know that they can request an accommodation or the types of accommodations that are available, according to another FEMA official, and therefore may not request them without being informed of their options. A FEMA disability integration advisor we interviewed said other FEMA staff may not have had the expertise to ask targeted questions to identify a disability and needed accommodations. In addition, FEMA's internal self-evaluation reported that there is no clear process for an individual with a disability to request an accommodation throughout FEMA's delivery of disaster assistance under IHP and individuals may not be aware of the process for making an accommodation request.[58]

FEMA has made efforts to add clarity to the disability-related questions. In September 2017, FEMA posted a video on social media in American Sign Language encouraging individuals with disabilities or accommodation requests to answer "yes" to the initial disability-related question. In 2018, FEMA revised the supplemental instructions that accompany the disability-related question asked during registration to add clarification about the disability-related information FEMA hopes to identify, according to FEMA officials.[59] FEMA established a work group to update the entire registration-intake questionnaire, and ODIC officials provided subject-matter expertise on the disability-related questions during

---

[58] The self-evaluation also found that FEMA demonstrated a trend of placing the burden of accessibility on the individuals requesting an accommodation rather than proactively incorporating accessibility mechanisms into its programs and services.

[59] The revised instructions read: *"We are asking this question to see if you or anyone in your household had a disability that affected mobility, vision, hearing, understanding others, or taking care of yourself before the disaster. We would also like to know if the disaster caused you or anyone in your household to not be able to perform daily living activities. Examples of daily living activities include, but are not limited to, bathing or showering, getting dressed, getting in and out of bed or a chair, walking, using the toilet, and eating. By answering yes to this question, we will be able to add more information about your disability and figure out the best way to assist you or others in your household."* FEMA provided training to Disaster Survivor Assistance staff focused on the new instructions, as well as tips for effectively assisting individuals with disabilities in answering the disability-related question.

Disaster Assistance 189

the effort. According to FEMA officials, FEMA is in the process of revising all the registration-intake questions in coordination with the Office of Management and Budget (OMB).[60] Ongoing efforts to provide additional instructions may help some survivors. However, until revisions to the disability-related questions are made that improve the clarity of the questions, FEMA could continue to miss opportunities to collect and share information that could assist more individuals with disabilities, such as by directly soliciting accommodation requests.

Individuals May Have Faced Challenges Receiving Needed Assistance Because FEMA Did Not Effectively Communicate Disability-Related Needs across Its Assistance Programs

Individuals may have faced challenges receiving necessary assistance because FEMA did not effectively track and communicate information about individuals' disability-related needs across its assistance programs after such needs were identified. Federal internal control standards call for effective internal communication, but FEMA officials told us the agency's process made it difficult to ensure that information was available to all FEMA staff who might need it.[61] FEMA's internal self-evaluation found that FEMA does not have comprehensive policies or procedures for processing and administering accommodation requests. When such requests are made, according to the evaluation, FEMA typically provides a one-time accommodation, but the request does not "follow" an individual throughout their interactions with FEMA. FEMA officials we interviewed explained that accommodation requests and disability-related information identified after registration-intake are recorded in a general "notes" section

---

[60] The Paperwork Reduction Act (PRA) provides that agencies may not conduct or sponsor the collection of information from 10 or more non-federal persons without first taking certain required steps, including allowing an opportunity for public comment and obtaining OMB approval, among other things. 44 U.S.C. § 3507. According to FEMA officials, FEMA cannot provide an estimate of when the process of updating the questions will be complete because it depends on OMB approval.

[61] GAO-14-704G.

190   *United States Government Accountability Office*

of a registrant's case file, which may include other information not related to a registrant's disability. Also, the electronic case file's design does not include an alert or flag to indicate an accommodation request, according to FEMA officials.

As different officials may handle different stages of FEMA's disaster assistance process, each official must read through the case file notes at various stages to see if accommodation requests or other disability-related needs were recorded. Officials acknowledged that disability-related needs recorded in the "notes" section can be easily overlooked as a case file is passed along to subsequent FEMA officials. An additional challenge is that FEMA's electronic case file prevents staff from updating a registrant's initial responses to the disability-related registration questions if a disability is identified or disclosed after the registration process. As a result, disability-related needs may not be communicated to FEMA staff who target outreach based in part on the answers to the registration-intake questions.

FEMA officials identified the home inspection as one area in the application process where challenges documenting and communicating information about a registrant's disability could result in reported disability-related needs and requested accommodations being inadvertently overlooked.[62] They said that home inspectors would ideally be informed of the registrant's disability-related needs or accommodation requests before the inspection. However, they explained that inspectors do not review registrants' full files and may only discover an individual's disability, such as a hearing or visual impairment, and any related accommodations the individual may need to participate in the inspection, when meeting with the individual for the first time at the inspection. Having an accommodation for an inspection could help individuals better answer the inspector's questions and identify the scope of the damage to their home. FEMA officials stated that a long-standing practice for individuals with hearing and visual impairment has been to leverage the help of friends and family

---

[62] FEMA home inspections are part of the process for verifying damage and loss for applicants for IHP assistance. Those who apply for FEMA assistance are typically contacted by a FEMA inspector or contractor to schedule an inspection.

Disaster Assistance 191

to communicate with the inspector. However, some people with disabilities may not have such support available.

DHS policy requires equal opportunity for people with disabilities served by its programs as well as effective communication to individuals who are deaf or hard of hearing or are blind or have low vision. However, a FEMA official said that some FEMA inspectors continued with inspections even after they discovered that an individual did not have the accommodation they requested—such as a sign language interpreter. Inspectors, who are typically contractors, may have done so because they were paid based on the number of inspections they complete and re-scheduling may be perceived as burdensome, according to the official.[63] Even when accommodation requests for inspections were identified, a limited pool of available inspectors and a lack of qualified sign-language interpreters can exacerbate delays in the inspection process, according to DHS's 2018 National Preparedness Report.[64]

Following the 2017 hurricanes, individuals who are older or have disabilities may have faced challenges with home inspections. Without information on a registrant's disability-related needs and requested accommodations, some home inspectors were not able to proactively address these needs when scheduling an inspection. Individuals with disabilities faced challenges scheduling and travelling to inspections following the 2017 hurricanes, according to local and nonprofit officials and a disability rights organization's report. Registrants who miss or cannot be contacted for an inspection may not be able to complete their application for FEMA assistance. FEMA requires inspectors to make a good faith effort to contact survivors.[65] However, people with disabilities,

---

[63] Another FEMA official reported in April 2019 that inspectors are no longer paid based on the number of inspections they complete.

[64] Department of Homeland Security, 2018 National Preparedness Report (Nov. 14, 2018). To meet the historically high needs for inspections in 2017, FEMA contracted additional inspectors to supplement existing inspections, according to the report. Still, FEMA experienced inspection staffing challenges that resulted in inspection backlogs across disaster affected areas. For example, due to inspection delays, on October 1, 2017, FEMA advised applicants in Texas that the inspection wait time may reach up to 45 days.

[65] According to the guidance FEMA provides to inspectors, to contact registrants, inspectors are required to try all available phone numbers, on different days, at different times of day, over a

192   *United States Government Accountability Office*

such as those who are deaf or hard of hearing, may not be able to respond to all forms of communication. A disability integration advisor described a case in which a FEMA team travelled to an elderly couple's home because FEMA staff could not contact them over the phone. The team learned that the couple could not hear the phone ring because they lost their hearing aids during the disaster. A disability integration advisor also stated that making information about individuals' accommodation needs readily available to FEMA staff could increase the success rate of contacting individuals with disabilities for services such as home inspections.

Challenges FEMA faced identifying and internally communicating disability-related needs, including the inspection-related challenges described above, may have contributed to some registrants not receiving FEMA's IHP assistance. For example, according to our analysis of FEMA data on registrations and awards for the 2017 hurricanes, 21 percent of registrants with reported disabilities who were denied assistance missed or could not be contacted for an inspection. For registrants with no reported disability whose application was denied, 19 percent missed or could not be contacted for an inspection.[66] In addition, FEMA staff stated that some individuals may not have received FEMA materials in accessible formats, such as large text for individuals with limited sight, and several nonprofit representatives in Florida reported that FEMA registration forms were not readily available in Braille. According to a FEMA Directive, alternate formats of disaster publications, such as documentation in Braille or large print, should not require special requests.[67] However, FEMA has not memorialized a process for obtaining these materials and requires individuals to request the alternate format each time they receive communication from FEMA, according to FEMA's internal self-evaluation.[68] Such challenges can prevent individuals with disabilities from

---

3-day period. Further, inspectors are instructed to use e-mail or texting if possible. Inspectors must also post a "Sorry I Missed You" note at the home with their name and contact info.

[66] This analysis does not assess compliance with any applicable non-discrimination or civil rights laws. In general, contacting individuals who have evacuated or who do not have electricity or cellular service can be challenging, according to a FEMA official.

[67] FEMA Directive 123-3, Printing, Duplicating and Copying (March, 2012).

[68] FEMA Directive 123-3, "Printing, Duplicating, and Copying" requires all disaster publications to be available with or without a request in Braille, large print, and electronic versions.

receiving FEMA benefits and fully engaging in FEMA programs and services. See appendix II for information on outcomes of FEMA applications for IHP assistance by disability status.

FEMA officials we interviewed said that communication of disability-related needs across the agency could be improved by creating an alert or field in case files to easily identify individuals' disabilities or accommodation requests. Such changes could also improve outcomes for this population, according to FEMA officials, and improve the agency's assistance to people with disabilities.

FEMA HAS TAKEN LIMITED STEPS TO EFFECTIVELY IMPLEMENT ITS NEW DISABILITY INTEGRATION APPROACH

FEMA Began Implementing its New Approach to Disability Integration without Articulating Objectives

In June 2018, FEMA began implementing changes to its disability integration approach in an effort to integrate disability competencies throughout the agency and among its partners. Specifically, one region established a new position for disability integration staff—the DIA intended to work with state and local emergency management agencies—in June 2018 and FEMA DIAs were deployed under the new model starting that same month. However, in an April 2018 memorandum to Regional Administrators outlining the staffing proposal, FEMA had not established objectives for the new approach and, as of December 2018, continued to lack a clear articulation of desired outcomes for its inclusive emergency management practices. Regarding FEMA's proposed change to add a new disability integration position to work directly with state or local emergency management agencies in each of the regions, ODIC officials

---

However, FEMA has only one disaster publication available in Braille, according to FEMA's internal self-evaluation.

told us that they provided a memorandum to all Regional Administrators outlining the proposal.[69] However, it does not include objectives for the new staffing approach. For example, the memorandum describes the role the new staff would play in building a culture of inclusive preparedness, including on accessibility-related evacuation, sheltering, and housing issues. However, it does not describe desired outcomes for FEMA's inclusive emergency management practices. ODIC officials acknowledged that Regional Administrators may interpret the new approach differently, and information provided by Regional Administrators shows that ODIC did not provide them with a set of objectives. Regarding deployments as part of its new disability integration approach, FEMA reported distributing to agency staff a document containing basic information on the approach, but this document does not outline objectives that agency leadership could use to gauge success. For example, it explains that FEMA is exploring ways to incorporate disability integration training into all deployable positions, but does not describe a set of outcomes that, if met, could show the training is effective.

Federal standards for internal control state that management should define objectives in specific and measurable terms and internally communicate the necessary information to achieve an agency's objectives.[70] Without a set of common objectives for FEMA's new disability integration approach, FEMA risks inconsistent application of its inclusive emergency management approach across its regions.

FEMA Has Not Communicated Key Information to Help Regions Implement the New Disability Integration Staffing Approach

ODIC officials have taken limited steps to communicate FEMA's new disability integration staffing approach in the regions to Regional

---

[69] ODIC officials said they first provided this memorandum to one Regional Administrator in April 2018 and later shared it with other Regional Administrators.

[70] GAO-14-704G.

Administrators and Regional Disability Integration Specialists (RDIS), who are critical to implementing these changes.[71] An ODIC official said that ODIC leadership began communicating basic information to Regional Administrators on its new approach to disability integration staffing in the regions through conference calls, informal conversations, and a memorandum starting in the fall of 2017.

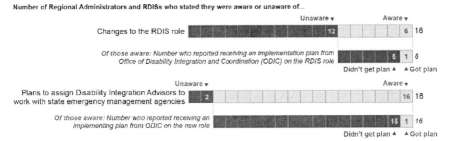

Source: GAO analysis of responses to structured questions from Federal Emergency Management Agency (FEMA) officials. | GAO-19-318.

Note: We collected responses to a set of structured questions from all ten FEMA Regional Administrators and eight RDISs in September 2018. Two of the 10 RDIS positions were vacant at the time we collected responses to our structured questions.

Figure 4. Differences of Awareness of FEMA's New Disability Integration Staffing Approach among All Regional Administrators and Regional Disability Integration Specialists (RDISs).

However, based on information Regional Administrators and RDIS reported to us in the fall of 2018, their awareness of FEMA's proposed changes varied (see Figure 4).[72] Further, most noted they had not received written implementation plans.

---

[71] As previously mentioned, Regional Disability Integration Specialists promote inclusive emergency management, support regional programmatic offices, and coordinate with state-assigned DIAs throughout FEMA's ten regions. Under the new staffing approach, FEMA proposed assigning new DIAs to each region to work with specific state or local emergency management agencies. In coordination with the RDIS, these DIAs would support state and local emergency managers by advising on inclusive emergency management principles and practices.

[72] We collected responses to a set of structured questions from all ten 10 FEMA Regional Administrators and eight RDIS in September 2018. Two of the 10 RDIS positions were vacant at the time we collected responses to our structured questions.

196   *United States Government Accountability Office*

Information Regional Administrators and RDISs reported to us in the fall of 2018 shows they lacked sufficient information from ODIC to better understand and consistently implement the new staffing approach nationwide. For instance, nearly all RDISs (seven of eight) reported not receiving information on their new responsibilities in working with DIAs assigned to work with state and local partners.[73] All 10 Regional Administrators reported that ODIC did not provide an implementation plan on the new DIA positions, and four said that improved communication from ODIC would be helpful to understand and implement the approach. While one RDIS reported receiving a description of revised responsibilities for DIA staff deployed to disaster locations, it did not address the new disability integration staffing approach in the regions. The memorandum that ODIC officials said they provided Regional Administrators on the new disability integration staff in the regions did not provide an implementation timeline or details, such as information on new staff roles and responsibilities, on what the new approach to disability integration should look like in the regions.

Further, ODIC has made changes to the new approach, acknowledging potential challenges with the plan to assign DIAs to each state. ODIC officials acknowledged that there may be differences in how the regions interpret the proposed changes to regional disability integration staff.

However, ODIC officials said they have no plans to provide an implementation timeline or additional details on staff responsibilities to facilitate implementation, stating that Regional Administrators have discretion on how to operationalize the new approach and do not need ODIC guidance.

In addition, ODIC officials have not distributed written practices on effective disability integration that could guide regional management in evaluating the performance of new regionally-based disability integration staff. All 10 Regional Administrators reported that they have not received guidance from ODIC on how to evaluate the performance of the DIAs assigned to develop state partnerships. Regional Administrators also said

---

[73] Two of the 10 RDIS positions were vacant at the time we conducted outreach.

they could use more information from ODIC, specifically on disability integration principles, which could help them evaluate staff performance by illustrating effective disability integration practices. ODIC officials reported that FEMA's human capital office is currently reviewing DIA position descriptions, but provided no estimate for when they would be available. ODIC officials reported verbally communicating their overarching philosophy on disability integration to Regional Administrators and RDIS, but said they have no plans to develop and distribute a written version. They explained that evaluating regional staff is the responsibility of the Regional Administrator, not ODIC.

Federal standards for internal control state that management should internally communicate the necessary information to achieve an agency's objectives.[74] It also states that management should design control activities that achieve objectives and respond to risks, and monitor and evaluate the results of those activities. Without a plan—including an implementation timeline and staff responsibilities—consistent with the objectives for the new approach to disability integration, FEMA risks inconsistent application of inclusive emergency management across its regions.

FEMA Has Taken Some Steps to Train Deployed Staff but No Longer Offers Comprehensive Training to External Partners

Training FEMA staff

To implement the new deployment model during disasters declared in 2018, FEMA officials from ODIC took some steps to train agency staff on disability integration to help ensure inclusive emergency management practices and reduce the reliance on DIAs in the field.

- Mandatory basic training on disability integration. In March 2018, the FEMA Administrator directed all FEMA staff to complete a 30- minute training on basic disability integration principles, such

---

[74] GAO-14-704G.

as how to support individuals with disabilities while working in the joint field office. FEMA officials characterized the training as an initial step in integrating disability competencies throughout the agency.
- "Just-in-time" training on disability integration. According to FEMA officials, instructors delivered a range of training on disability integration in emergency management to agency staff deploying to disaster locations. For example, one of these trainings focused on communication and etiquette basics when interacting with individuals with disabilities. Another trained staff on how to ask individuals about their disabilities for the registration intake form.

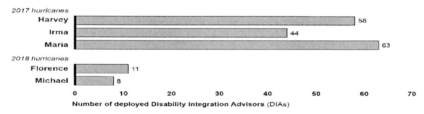

Source: GAO analysis of Federal Emergency Management Agency (FEMA) DIA deployment data. | GAO-19-318.

Note: This data reflects the peak number of deployed DIAs to select disaster locations in 2017 and 2018. FEMA shifted DIAs among those locations, depending on demands, during response and recovery efforts. In 2017, DIAs directly assisted hurricane survivors with disabilities in Texas, Florida, Puerto Rico, and the U.S. Virgin Islands. In 2018, under a new model, DIAs advised FEMA field leadership in North Carolina, South Carolina, Georgia, and Florida on disability integration practices.

Figure 5. Peak Number of Deployed FEMA Disability Integration Advisors in 2017 and 2018 Hurricane Locations.

After implementing its new disability integration staffing deployment model in June 2018, FEMA reduced the number of DIAs deployed to disaster locations (see Figure 6).

As part of the shift in responsibilities, FEMA officials reported that DIAs deployed to 2018 disasters provided none of the disability integration training to other FEMA staff.[75] According to FEMA officials, these trainings instead were delivered by FEMA training staff on an as-needed basis. While officials told us in November 2018 that they planned to evaluate the trainings through surveys, they had not yet done so as of December 2018.

Although FEMA officials emphasized the need to integrate disability competencies throughout its programmatic offices and deployable staff, they do not have written plans—including milestones, performance measures, or a plan for monitoring performance—for developing new comprehensive training for all staff beyond the basic and just-in-time training currently available. In July 2018, FEMA hired a new Program and Policy Branch Chief for ODIC who, according to ODIC officials, will review, assess, and recommend ways to incorporate disability competencies into training for all FEMA staff. ODIC officials said that program and field staff will help identify training gaps to inform the chief's recommendations.[76] However, officials reported no timeline for making these recommendations. According to one official, FEMA will develop new training as needed. The former FEMA Administrator, under whom ODIC was created, noted that significant investment would be required to fully integrate disability competencies throughout the agency. He explained that it will take time for FEMA to comprehensively train staff on disability competencies and that DIAs are necessary for filling in gaps in the meantime. Current FEMA officials also acknowledged that more training will be needed to accomplish their goals related to inclusive emergency management. However, they noted that building a disability integration curriculum will take time and funding, explaining that FEMA's training unit may have competing priorities and the agency's 2019 and future budgets are unknown.

---

[75] Under the new model, DIAs are responsible for advising field leadership and program staff on inclusive emergency management practices, and other deployed FEMA staff are responsible for directly assisting individuals with disabilities. According to ODIC officials, this responsibility has always existed for all FEMA staff, but DIAs have increasingly taken over more of the responsibility since 2010 when ODIC was created.

[76] ODIC officials said that they are also developing a strategic plan that will address state and local capacity building.

200 *United States Government Accountability Office*

FEMA's August 2017 internal self-evaluation reported that its staff generally lacked adequate disability-related training and education. During interviews in all four disaster sites, officials and others we interviewed, including local officials, disability advocates, and survivors, reported that FEMA staff did not always effectively communicate with and assist individuals who are older or have disabilities in completing the online registration-intake form. Federal standards for internal control state that management should identify, analyze, and respond to significant changes that could impact the internal control system.[77] Additionally, leading practices identified in the Program Management Institute's The Standard for Program Management call for agencies to develop meaningful measures to monitor program performance and to track the accomplishment of the program's goals and objectives.[78] We have previously recommended that FEMA establish and use goals, milestones, and performance measures to monitor program performance designed to help FEMA staff determine their readiness to respond to disasters.[79] FEMA's Strategic Plan 2018-2022 states that FEMA will develop targeted solutions to close identified competency gaps. Developing a plan for delivering disability integration training to all FEMA staff that includes milestones and performance measures, and outlines how performance will be monitored, would better position FEMA to provide training to all staff that achieves its intended goals. As a result of such training, deployed staff will be better equipped to identify and assist individuals who are older or have disabilities.

Training for Nonfederal Partners

The smaller number of DIAs deployed, as well as their shift away from providing direct assistance to individuals with disabilities, may result in nonfederal partners, such as state, territorial, and local emergency

---

[77] GAO-14-704G.

[78] Project Management Institute, Inc., The Standard for Program Management ®, Third Edition (Newton Square, PA: 2013). The Standard for Program Management ® describes, among other things, how resource planning; goals, milestones, and performance measures; and program monitoring and reporting are good practices that can enhance management for most programs.

[79] GAO-15-781.

managers, providing more direct assistance to these individuals than they did previously. However, according to officials, FEMA stopped offering its comprehensive introductory course, "Integrating Access and Functional Needs into Emergency Planning," on disability integration to its nonfederal partners in September 2017. We previously reported that this 2-day training included substantial information on incorporating the needs of people with disabilities in emergency planning.[80] For example, the training included a module for emergency managers on the importance of preparing shelters with equipment to accommodate individuals with disabilities. We also reported that the demand for the training exceeded FEMA's delivery capacity.[81] However, ODIC officials determined that the course did not provide actionable training to emergency management partners and other stakeholders to meet the needs of individuals with disabilities.[82] FEMA officials said they have plans to replace the course with new training, which is currently under development, but provided no timeline for doing so.

During interviews in Florida, Texas, Puerto Rico, and the U.S. Virgin Islands, officials from four disability organizations told us that state and local emergency managers could use additional disability-related training. Specifically, these stakeholders said that individuals with disabilities reported experiences where emergency managers were insensitive to their needs and disability-related accommodation requests. For example, disability advocates and survivors in Florida, Puerto Rico, Texas, and the U.S. Virgin Islands described shelter-related accessibility challenges, including access to bathrooms. In addition, three deployed DIAs we interviewed in Florida and Puerto Rico reported that local emergency

---

[80] GAO-17-200.

[81] To address this challenge, we recommended that FEMA establish written procedures for involving ODIC in regional activities; set goals for disseminating disability integration training; and evaluate alternative delivery methods for the training. FEMA agreed with the recommendations, and has efforts underway to address them. See GAO-17-200.

[82] According to FEMA officials, staff in some FEMA regions have initiated advisory groups as a way to train and engage disability partners and local emergency managers. FEMA officials said that RDISs in six regions facilitate relationships between local emergency management agencies and disability stakeholders using advisory groups. FEMA has used advisory groups to help jurisdictions build inclusive emergency management practices through collaboration and partnership between disability stakeholders and emergency managers.

managers have requested structured training on disability integration practices to better prepare their communities.

Federal standards for internal control state that management should externally communicate the necessary quality information to effectively implement the agency's objectives and, under the Post-Katrina Act, one of the Disability Coordinator's responsibilities is ensuring the development of training materials for emergency managers.[83] A timeline for completing the development of new training for state and local emergency managers would help ensure such training is offered and partners are provided with timely information on inclusive emergency management practices.

CONCLUSION

The unprecedented 2017 hurricane season tested emergency managers' capacity for responding to the sometimes complex needs of individuals who are older or have disabilities in Florida, Puerto Rico, Texas, and the U.S. Virgin Islands. Some challenges—such as debilitated transportation infrastructure in the U.S. Virgin Islands and Puerto Rico—proved overwhelming for local and state officials. FEMA tried to mitigate these challenges with its disability integration staff, who aimed to ensure that individuals with disabilities received appropriate assistance.

Nevertheless, the difficulties individuals experienced in attempting to register for FEMA assistance, as well as local and nonprofit partners' difficulties gaining access to registration information, may have limited the assistance older individuals and those with disabilities received. Confusion about the registration questions and the lack of communication across FEMA programs may have led to FEMA overlooking individuals with disability-related assistance needs. As a result, FEMA and its partners may have missed opportunities to collect and share information that would have allowed them to better assist individuals who are older or have disabilities.

---

[83] GAO-14-704G; 6 U.S.C. § 321b(b)(5).

Disaster Assistance 203

FEMA's new approach to decentralize and distribute responsibilities across the agency for providing disaster assistance to individuals who have disabilities was not broadly, clearly, or consistently articulated. FEMA may not be able to effectively implement its new disability integration approach without objectives or a plan that includes a timeline and staff responsibilities. These elements are critical for ensuring that all FEMA staff are consistently delivering appropriate services to disaster survivors with disability-related needs. These changes are also being implemented before staff have been fully trained. Without a plan for delivering training on disability integration to FEMA staff that includes milestones and performance measures, and that outlines how performance should be monitored, the agency may be ill-prepared to identify and address challenges individuals who have disabilities, including those who are older, face recovering from disaster. Furthermore, without a timeline for completing the training FEMA has begun developing for state and local emergency management partners on inclusive emergency management practices, there is a risk partners will not have timely access to information that would benefit individuals most vulnerable to the effects of disasters.

RECOMMENDATIONS FOR EXECUTIVE ACTION

We are making the following seven recommendations to the FEMA Administrator:

- The FEMA Administrator should develop and publicize guidance for partners working to assist individuals who are older or have disabilities for requesting data and working with FEMA staff throughout the data sharing process to obtain Individual Assistance data, as appropriate. (Recommendation 1)
- The FEMA Administrator should implement new registration-intake questions that improve FEMA's ability to identify and address survivors' disability-related needs by, for example,

204 *United States Government Accountability Office*

directly soliciting survivors' accommodation requests. (Recommendation 2)

- The FEMA Administrator should improve communication of registrants' disability-related information across FEMA programs, such as by developing an alert within survivor files that indicates an accommodation request. (Recommendation 3)
- The FEMA Administrator should establish and disseminate a set of objectives for FEMA's new disability integration approach. (Recommendation 4)
- The FEMA Administrator should communicate to Regional Administrators and Regional Disability Integration Specialists a written plan for implementing its new disability integration staffing approach, consistent with the objectives established for disability integration. Such a plan should include an implementation timeline and details on staff responsibilities, which regions could use to evaluate staff performance. (Recommendation 5)
- The FEMA Administrator should develop a plan for delivering training to FEMA staff that promotes competency in disability awareness. The plan should include milestones and performance measures, and outline how performance will be monitored. (Recommendation 6)
- The FEMA Administrator should develop a timeline for completing the development of new disability-related training the agency can offer to its partners that incorporates the needs of individuals with disabilities into disaster preparedness, response, and recovery operations. (Recommendation 7)

AGENCY COMMENTS AND OUR EVALUATION

We provided a draft of this chapter to DHS for review and comment. DHS provided written comments, which are reproduced in appendix III and described below, and concurred with recommendations 1, 2, and 4-7. The comment letter generally described steps FEMA plans to take, or is in the process of taking, to address all recommendations. As discussed further below, DHS did not concur with recommendation 3 and, for recommendation 6, described plans to improve disability competencies among FEMA staff that does not include training. DHS also provided technical comments, which we incorporated as appropriate.

DHS did not concur with recommendation 3 to improve communication of registrants' disability-related information across FEMA programs. Specifically, DHS noted that FEMA lacks specific funding to augment the legacy data systems that capture and communicate registration information. DHS further noted that FEMA began a long-term initiative in April 2017 to improve data management and exchange, and improve overall data quality and standardization. FEMA expects the initiative to include the development of a modern, cloud-based data storage system with a data analytics platform that will allow analysts, decision makers, and stakeholders more ready access to FEMA data. After the completion of this initiative, FEMA expects that efforts to share and flag specific disability-related data will be much easier.

We acknowledge FEMA's concerns about using resources to change legacy systems when it has existing plans to replace those systems. However, the recommendation was not solely focused on system changes, although that is an example of a way to help improve communication. There are other cost-effective ways that are likely to improve communication of registrants' disability-related information prior to implementing the system upgrades. For example, as noted in the report, FEMA officials handling different stages of the disaster assistance process may overlook disability-related needs recorded in the case file notes. FEMA could revise its guidance to remind program officials to review the notes to identify whether there is a record of any such needs. As FEMA moves ahead with its data improvement initiatives, we encourage it to consider and ultimately implement technology changes, such as developing an alert within files that indicates an accommodation request. Such

206   *United States Government Accountability Office*

improvements would be consistent with the recommendation and help improve communication across FEMA programs.

With respect to recommendation 6 related to training FEMA staff on disability competencies, DHS concurred. The letter stated that FEMA's ODIC is developing a plan to include a disability integration competency in the position task books for all deployable staff, rather than through training. Position task books outline the required activities, tasks, and behaviors for each job, and serve as a record for task completion. The plan will also include:

- a communications and outreach plan;
- milestones for measuring the effectiveness of the integration of this competency across the agency; and
- a monitoring plan and milestones to measure the overall integration of this competency across the deployable workforce.

We continue to believe that FEMA should develop a plan that includes how it will deliver training to promote competency in disability awareness among its staff. The plan for delivering such training should include milestones, performance measures, and how performance will be monitored.

As agreed with your offices, unless you publicly announce its contents earlier, we plan no further distribution until 30 days from its issue date. At that time, we will send copies of this chapter to the appropriate congressional committees, the Secretary of Homeland Security, and other interested parties.

Elizabeth Curda
Director, Education, Workforce, and Income Security Issues

List of Requesters

The Honorable Michael Enzi
Chairman
Committee on the Budget
United States Senate

The Honorable Ron Johnson
Chairman

The Honorable Gary C. Peters
Ranking Member
Committee on Homeland Security and Governmental Affairs
United States Senate

The Honorable Marco Rubio
Chairman
Committee on Small Business and Entrepreneurship
United States Senate

The Honorable Susan Collins
Chairman

The Honorable Robert Casey
Ranking Member
Special Committee on Aging
United States Senate

The Honorable Rand Paul, MD
Chairman
Subcommittee on Federal Spending, Oversight and Emergency Management
Committee on Homeland Security and Governmental Affairs
United States Senate

The Honorable Maxine Waters
Chairwoman
Committee on Financial Services
House of Representatives

The Honorable Bennie Thompson
Chairman
Committee on Homeland Security
House of Representatives

The Honorable Elijah Cummings
Chairman

The Honorable Jim Jordan
Ranking Member
Committee on Oversight and Reform
House of Representatives

The Honorable Nydia Velázquez
Chairwoman
Committee on Small Business
House of Representatives

The Honorable Peter DeFazio
Chairman

The Honorable Samuel "Sam" Graves
Ranking Member
Committee on Transportation and Infrastructure
House of Representatives

The Honorable Al Green
Chairman
Subcommittee on Oversight and Investigations

Committee on Financial Services
House of Representatives

The Honorable Sean Duffy
Ranking Member
Subcommittee on Housing, Community Development, and Insurance
Committee on Financial Services
House of Representatives

The Honorable Emanuel Cleaver, II
House of Representatives

The Honorable Michael McCaul
House of Representatives

The Honorable Gary Palmer
House of Representatives

The Honorable Ann Wagner
House of Representatives

APPENDIX I: OBJECTIVES, SCOPE, AND METHODOLOGY

This appendix discusses in detail our methodology for examining 1) challenges Federal Emergency Management Agency (FEMA) partners reported in providing disaster assistance to individuals who are older or have disabilities; 2) challenges faced by such individuals in accessing FEMA's disaster assistance programs, and actions FEMA has taken to address such challenges; and 3) the extent to which FEMA has planned for and taken steps to implement its new approach to disability integration. The focus of this chapter is on the three near-sequential hurricanes— Hurricanes Harvey, Irma, and Maria—that made landfall in 2017 in

210   *United States Government Accountability Office*

primarily four geographic areas—Florida, Puerto Rico, Texas, and the U.S. Virgin Islands.[84]

Primarily to address our first objective, we visited Florida, Puerto Rico, Texas, and the U.S. Virgin Islands in June and July 2018. At each location we interviewed state or territory emergency managers, public health and human services officials, and representatives of nonprofit disability organizations. For example, we interviewed representatives of Centers for Independent Living in all four locations and organizations that advocate for the civil rights of people with disabilities in Texas and Florida. We also interviewed local emergency managers in Texas and Florida in counties that were affected by Hurricanes Harvey and Irma. To learn first-hand accounts of disaster-related challenges faced by individuals in Puerto Rico and the U.S. Virgin Islands who are older or have disabilities, staff from the Centers for Independent Living in those locations invited us to interview a collective total of 16 of their regular program participants.

We also interviewed representatives of national organizations, selected for their focus on providing assistance to disaster survivors who are older or have disabilities, including AARP, The Partnership for Inclusive Disaster Strategies—a coalition focused on inclusive disaster planning and emergency preparedness for people with disabilities—and Portlight Inclusive Disaster Strategies, Inc. In addition, we interviewed officials from other relevant federal agencies and offices, including the Department of Homeland Security's (DHS) Office for Civil Rights and Civil Liberties (CRCL) and the National Council on Disability. To supplement information we obtained from interviews, we reviewed summaries of eight public listening sessions published by CRCL and co-hosted with FEMA across the four disaster locations between February 2018 and May 2018. The purpose of these sessions, according to CRCL, was to hear about concerns and experiences related to the impact of the disasters on individuals with disabilities. While the perspectives of officials and

---

[84] We also spoke with FEMA officials and disability advocates and collected limited data on FEMA's response to hurricane Florence and Hurricane Michael that made landfall in 2018 in North Carolina and Florida.

Disaster Assistance 211

stakeholders we interviewed, as well as those expressed during the public listening sessions, are not generalizable, they provide valuable insights into the federal response to the 2017 disasters.

To address our second objective, we obtained and analyzed summary data from FEMA's National Emergency Management Information System (NEMIS)—a database used to track disaster data—on FEMA registrations and awards for disaster assistance for the hurricanes included in our review. FEMA also provided data for registrations submitted by households with residents who are older and households with residents who reported disabilities. Registrations were included in the category of older survivor if the applicant or co-applicant was aged 65 or older. Registrations were included in the disability category if the applicant answered "yes" to the to the registration-intake question: "Did you, your co-applicant, or any dependents have help or support doing things like walking, seeing, hearing, or taking care of yourself before the disaster and have you lost that help or support because of the disaster?" These data reflect the status of FEMA registrations as of October 18, 2018.

We also obtained and analyzed summary data from FEMA's internal and external call centers that operate its helpline, including the number of calls that were answered and the average wait times for answered calls for a given day.[85] We focused on the busiest period of incoming calls for the call centers following the 2017 hurricanes—August 2017 to October 2017. To assess the reliability of FEMA's summary data, we interviewed officials at FEMA headquarters about the quality of the data; reviewed existing information about the data systems; and conducted checks for out of range or logically inaccurate data and comparisons to publicly available summary data. The demographic data in NEMIS, such as disability-related information, are largely self-reported by applicants, and FEMA does not independently verify all of the data it collects.

Further, as our report identifies, the number of individuals who answer yes to the disability-related question on FEMA's registration form may not be a reliable estimate for the number of individuals who have a disability.

---

[85] Individuals can call FEMA's helpline to complete their registration for FEMA assistance, ask questions about the application process, or check on the status of an existing application.

We determined that the data were sufficiently reliable for the purpose of providing information on the number and characteristics of registrations for assistance FEMA collects and the number of calls to FEMA's call centers.

To address our third objective, we compared staffing levels before and after FEMA implemented its new disability integration approach by obtaining and analyzing the number of deployed disability integration officials in response to the 2017 hurricanes, and the 2018 Hurricanes Florence and Michael. To assess the reliability of these data we reviewed recent GAO work that assessed the reliability of FEMA's workforce data from the same data source and reviewed the data for obvious errors and omissions.[86] We determined that the data were sufficiently reliable to provide information on the deployed FEMA workforce in response to recent hurricanes. To further assess FEMA's planning and implementation of its new approach to disability integration, we obtained and analyzed responses to structured questions from officials in FEMA's ten regions.

To address all three objectives, we reviewed relevant federal laws and regulations; relevant agency documents; and reports by federal agencies and nongovernmental organizations. We analyzed FEMA policies, procedures, guidance, and memoranda specific to FEMA's Individuals and Households Program and disability integration, including FEMA's plan to implement its new approach to disability integration and its data sharing policy. We assessed FEMA's actions against goals and objectives in FEMA's 2018-2022 Strategic Plan; DHS policy for ensuring nondiscrimination for individuals with disabilities; and federal standards for internal control related to effective internal and external communication, using quality information to achieve objectives, and defining objectives in measurable terms.

We reviewed federal disaster-related frameworks and reports, including the National Response Framework, the 2018 National Preparedness Report, the 2017 Hurricane Season FEMA After-Action Report. We also reviewed FEMA's unpublished self-evaluation of its

---

[86] GAO-18-472.

Disaster Assistance 213

policies and practices, which sought to determine how effectively FEMA provides equal physical, program, and communication access to people with disabilities.[87] We also reviewed relevant information from our prior reports on FEMA's work,[88] as well as an after-action report from The Partnership for Inclusive Disaster Strategies—a disability rights organization.[89] We did not independently assess whether any programs or activities conducted by FEMA or its partners during the period covered by our review complied with applicable non-discrimination or civil rights laws.

To address all three objectives, we also interviewed FEMA officials to discuss FEMA's process for sharing disaster assistance data, challenges individuals may have faced accessing FEMA's disaster assistance programs, and FEMA's new approach to disability integration. We interviewed staff from FEMA headquarters, including officials from the Office of Disability Integration and Coordination, Office of External Affairs, and officials who administer the Individual Assistance Program. We also met with FEMA staff focused on assisting individuals with disabilities and deployed to each disaster location, including staff from the FEMA regional offices where each disaster occurred. Finally, we

---

[87] In 2013, the Department of Homeland Security (DHS) issued Directive 065-01, Nondiscrimination for Individuals with Disabilities in DHS-Conducted Programs and Activities (Non-Employment), which included a requirement for DHS components—including FEMA—to (1) conduct a self-evaluation of their programs and activities to identify any barriers to access and gaps in existing component policies or procedures for providing reasonable accommodations; and (2) develop a plan that addresses any identified barriers and documents the components' disability policies. FEMA completed its self-evaluation in August 2017 and plans to issue a plan for addressing issues identified in the self-assessment in 2019. GAO received the self-evaluation and a draft implementation plan in December 2018, after our site visits and the majority of our audit work was completed.

[88] GAO, *2017 Hurricanes and Wildfires: Initial Observations on the Federal Response and Key Recovery Challenges,* GAO-18-472, (Washington, D.C.: Sept. 4, 2018) and GAO, *Federal Disaster Assistance: FEMA's Progress in Aiding Individuals with Disabilities Could Be Further Enhanced,* GAO-17-200 (Washington, D.C.: Feb. 7, 2017).

[89] This report incorporated the perspectives of stakeholders who were working in disaster-impacted communities and thousands of callers to a hotline established by The Partnership for Inclusive Disaster Strategies to address the needs of survivors of the 2017 hurricanes. The organization's CEO is a former director of FEMA's Office of Disability Integration and Coordination. The Partnership for Inclusive Disaster Strategies, *Getting It Wrong: An Indictment with a Blueprint for Getting It Right. Disability Rights, Obligations and Responsibilities Before, During and After Disasters* (May 2018).

214 *United States Government Accountability Office*

interviewed former FEMA officials, including a previous FEMA Administrator.

We conducted this performance audit from April 2018 to May 2019, in accordance with generally accepted government auditing standards.

Those standards require that we plan and perform the audit to obtain sufficient, appropriate evidence to provide a reasonable basis for our findings and conclusions based on our audit objectives. We believe that the evidence obtained provides a reasonable basis for our findings and conclusions based on our audit objectives.

APPENDIX II: OUTCOMES OF APPLICATIONS FOR FEDERAL EMERGENCY MANAGEMENT AGENCY (FEMA) FINANCIAL ASSISTANCE FOR 2017 HURRICANES

According to our analysis of FEMA data, registrants who reported a disability during registration-intake had a higher percentage of denial for FEMA IHP assistance, a lower percentage of submitting an appeal, and a higher percentage of denial on appeal compared to registrants who did not report a disability (see Figure 6).[90] Individuals with disabilities have unique challenges, according to FEMA officials. These challenges may result in increased rates of denial for disaster assistance.

For example, people with disabilities, particularly those with complex medical needs, may require extensive documentation to prove losses and obtaining medical records may be challenging after a storm.

---

[90] Our analysis did not assess compliance with any applicable non-discrimination or civil rights laws. Registrants 65 and older had a lower percentage of denials for FEMA assistance, a higher percentage of submitting an appeal, and a lower percentage of denial for appeal compared to registrants who were under 65. Specifically, 36.1 percent of registrants 65 and older (219,954) were initially denied assistance; 18.3% of those denials (40,245) were appealed; and 74.9 percent of those appeals (30,153) were denied. In comparison, 38.9 percent of registrants under 65 (1.2 million) were initially denied assistance; 8.2 percent of those denials (95,486) were appealed; and 76.0 percent of those appeals (72,579) were denied.

Disaster Assistance 215

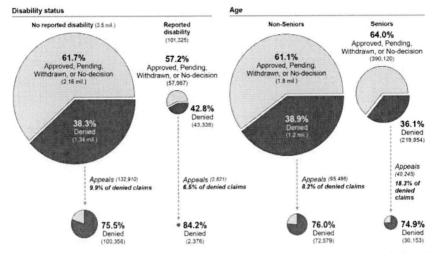

Source: GAO analysis of Federal Emergency Management Agency (FEMA) data. | GAO-19-318.

Note: The data include registrations that were referred for Federal Emergency Management Agency's (FEMA) Individuals and Households Program in Florida, Puerto Rico, Texas, and the U.S. Virgin Islands as a result of Hurricane Harvey, Irma, and Maria in 2017. Some registrations, such as those indicating no damage, are not referred for assistance, according to FEMA. The "reported disability" population includes individuals who answered 'yes' to the registration-intake question: "Did you, your co-applicant, or any dependents have help or support doing things like walking, seeing, hearing, or taking care of yourself before the disaster and have you lost that help or support because of the disaster?" This analysis does not assess compliance with any applicable non-discrimination or civil rights laws.

Figure 6. Number of Denied Applications and Appeals for Federal Emergency Management Agency (FEMA) Financial Assistance for 2017 Hurricanes by Disability Status and Age, as of October 18, 2018.

Appendix III: Comments from the Department of Homeland Security

April 18, 2019

Elizabeth Curda, Director
Education, Workforce, Income Security
U.S. Government Accountability Office
441 G Street, NW
Washington, DC 20548

Re: Management Response to Draft Report GAO-19-318, "DISASTER ASSISTANCE: FEMA Action Needed to Better Support Individuals Who Are Older or Have Disabilities"

Dear Ms. Curda:

Thank you for the opportunity to review and comment on the subject draft report. The U.S. Department of Homeland Security (DHS) appreciates the U.S. Government Accountability Office's (GAO) work in planning and conducting its review and issuing this report.

The Department is pleased to note GAO's recognition of the unprecedented nature of the 2017 hurricane season and Federal Emergency Management Agency (FEMA) efforts to mitigate challenges associated with responding to the sometimes complex needs of individuals who are older or have disabilities. This included efforts to implement an enterprise level approach to disability integration and the Agency's strategy to incorporate a disability integration competency across all FEMA programs.

More specifically, following hurricanes Harvey, Irma, and Maria and the wildfires in California, the DHS Office for Civil Rights and Civil Liberties (CRCL) and FEMA Office of Equal Rights (OER) took a variety of steps to ensure that FEMA as well as state, local, and territorial agencies that receive federal assistance comply with civil rights laws, including Section 504 of the Rehabilitation Act of 1973, in disaster preparedness, response, and recovery programs and activities.

For example, together with the U.S. Department of Justice Civil Rights Division, these offices reminded state and local emergency management agencies of their civil rights responsibilities in a letter dated October 4, 2018. The letter offers technical assistance, outreach, and training to help state and local emergency management agencies

Disaster Assistance

understand and meet their responsibilities for civil rights compliance when they receive federal funding.

In addition, in March of 2019, DHS Officer for CRCL sent a letter with recommendations to the states and territories where DHS CRCL, OER, and FEMA's Office of Disability and Integration Coordination (ODIC) held the listening sessions. These recommendations, which will be distributed more broadly to emergency management agencies at all levels, are intended to improve the delivery of disaster assistance to persons with disabilities. States and territories, like its counties and any other subrecipients of DHS financial assistance, must ensure that persons with disabilities have equal access to federally funded programs and activities as well as the offer to provide technical assistance to support civil rights compliance.

Recipients and the public alike have access to these recommendations and other federal resources on the DHS webpage, Civil Rights in Emergencies and Disasters, found at https://www.dhs.gov/civil-rights-emergencies-and-disasters. CRCL will also be working to raise awareness about the recommendations, and practical ways to implement these, by engaging with national homeland security and emergency management associations.

At the national and regional levels, FEMA convenes disability stakeholders from the public, private, and non-profit sectors to identify strategies, create tools and develop solutions to eliminate gaps in FEMA programs and services to meet the needs of people with disabilities before, during and after disasters. To ensure that all communities are being served during disasters, FEMA has highlighted these initiatives in its 2018-2022 Strategic Plan. FEMA remains committed to incorporating equal access for, and inclusion of, individuals with disabilities in all phases of emergency management.

The draft report contained seven recommendations with which the Department concurs on six and non-concurs on one. Attached find our detailed response to each recommendation. Technical comments were previously provided under separate cover.

Again, thank you for the opportunity to review and comment on this draft report. Please feel free to contact me if you have any questions. We look forward to working with you again in the future.

Sincerely,

JIM H. CRUMPACKER, CIA, CFE
Director
Departmental GAO-OIG Liaison Office

218   *United States Government Accountability Office*

**Attachment: Management Response to Recommendations
Contained in GAO-19-318**

GAO recommended that the FEMA Administrator:

Recommendation 1: Develop and publicize guidance for partners working to assist individuals who are older or have disabilities for requesting data and working with FEMA staff throughout the data sharing process to obtain Individual Assistance data.

Response: Concur. FEMA is currently developing new, enhanced templates, policies and guidance for FEMA field staff to facilitate data sharing with States and other partners. FEMA's Recovery Reporting and Analytics Division (RRAD) will be providing a full-service capability to FEMA personnel to support them throughout the data sharing process. In addition, FEMA is expanding its Open FEMA datasets to provide more autonomous, self-service sharing of aggregated data to fulfill partner data needs when personally identifiable information data is not required. FEMA will also be publishing data sharing guidelines to the FEMA.gov internet site soon describing all the ways to obtain FEMA data to best serve individuals that are older or who have disabilities. Estimated Completion Date (ECD): March 31, 2020.

Recommendation 2: Implement new registration intake questions that improve FEMA's ability to identify and address survivors' disability-related needs by, for example, directly soliciting survivors' accommodation requests.

Response: Concur. FEMA's Individuals and Households Program (IHP) Branch will continue to coordinate with FEMA's ODIC and OER to improve the registration intake questions related to the identification of disaster-related loss to disability-based equipment and services. In addition to that coordination, IHP and ODIC will engage with RRAD to assess the demographic factors that represent the percentage of survivors positively answering registration questions regarding the need for assistive support. The data analysis will also assess whether FEMA's recent addition of help text language within registration intake has improved the number of survivors who positively answer registration questions regarding the need for assistive support.

Based on the analysis, FEMA will determine if the data suggest changes to the existing registration intake questions are necessary or whether new registration intake questions should be developed to improve FEMA's ability to identify and address survivors' accessibility-related needs. FEMA will coordinate any changes to the registration intake questions with the U.S. Office of Management and Budget, as required. ECD: March 31, 2020.

Disaster Assistance

219

Recommendation 3: Improve communication of registrants' disability-related information across FEMA programs, such as by developing an alert within survivor files that indicates an accommodation request.

Response: Non-concur. FEMA currently lacks specific funding resources to augment the legacy systems that capture and communicate registration information. However, FEMA is working to improve the Agency's foundational data management and exchange capabilities through the Enterprise Data and Analytics Modernization Initiative (EDAMI), started in April 2017. This long term program is maturing data management across the Agency to improve overall data quality and standardization. EDAMI also includes the development of the FEMA Data Exchange (FEMADex), a modern, cloud-based data-storage solution with an advanced data analytics platform that will allow analysts, decision makers, and stakeholders more ready access to FEMA data. Efforts to share and flag specific disability-related data will be much easier once the Agency has improved these foundational data management and exchange capabilities.

FEMA requests that the GAO consider this recommendation resolved and closed.

Recommendation 4: Establish and disseminate a set of objectives for FEMA's new disability integration approach.

Response: Concur. FEMA's ODIC is in the process of developing its 2019-2022 Strategic Plan. This plan establishes the objectives, milestones, and tactics to implement FEMA's enterprise level approach to disability integration. ODIC will coordinate with OER to clarify roles and responsibilities between these offices with certain overlapping responsibilities for disability integration, including training.

The development of the Strategic Plan is informed by:

(a) two Partner Strategy Sessions conducted in Washington, DC with state, local, tribal, and territorial partners as well as Non-Government Organizations and private sector partners in September and October of 2018;

(b) interviews with Senior FEMA Leadership in Headquarters, across FEMA's programs that provide recovery resources, Regional Administrators, and Field Leadership begun in 2017 and ongoing as of the drafting of this management response letter; and

(c) an ongoing engagement with FEMA's Continuous Improvement Program.

The Plan is aligned with FEMA's 2018-2022 Strategic Plan, integrating lessons learned through deployments in 2017 and 2018. ODIC is concurrently developing a communication plan to socialize the Strategic Plan throughout the Agency. ECD: June 30, 2019.

Recommendation 5: Communicate to Regional Administrators and Regional Disability Integration Specialists a written plan for implementing its new disability integration

220 *United States Government Accountability Office*

staffing approach, consistent with the objectives established for disability integration. Such a plan should include an implementation timeline and details on staff responsibilities, which regions could use to evaluate staff performance.

Response: Concur. FEMA's ODIC is in the process of developing a project plan for the implementation of the proposed staffing approach in the Regions. This project plan will include an implementation timeline, detailed roles and responsibilities for the proposed Disability Integration Advisor positions, a force structure for each Region, and recommendations to the Regional Administrator for establishing performance metrics for the new positions that support the Agency's overall enterprise approach to disability integration. This project plan will be completed and presented to the Administrator for review and concurrence. ECD: June 30, 2019.

Recommendation 6: Develop a plan for delivering training to FEMA staff that promotes competency in disability awareness. The plan should include milestones and performance measures, and outline how performance will be monitored.

Response: Concur. FEMA's ODIC is developing a plan to include a disability integration competency in the Position Task Books for all Incident Management deployable titles in the Agency. This plan will include a communications and outreach plan to socialize the enterprise level approach to disability integration, milestones for measuring the effectiveness of the integration of this competency across the agency, and a monitoring plan and milestones to measure the overall integration of this competency across the deployable workforce. This plan will be presented to FEMA's Field Operations Directorate for concurrence before being presented to the Administrator for approval. ECD: September 30, 2019.

Recommendation 7: Develop a timeline for completing the development of new disability-related training the agency can offer to its partners that incorporates the needs of individuals with disabilities into disaster preparedness, response, and recovery operations.

Response: Concur. FEMA's ODIC has developed a set of courses of action for updating E/L 0197: "Integrating Access and Functional Needs into Emergency Planning," and an implementation decision by the Acting FEMA Administrator is pending. ECD: June 30, 2019.

INDEX

A

accommodation, 149, 182, 183, 185, 188, 189, 190, 191, 192, 193, 201, 204, 206

agencies, x, 37, 38, 40, 41, 56, 58, 59, 75, 78, 86, 91, 122, 126, 138, 139, 140, 143, 153, 157, 159, 160, 163, 171, 173, 174, 179, 182, 186, 189, 193, 195, 200, 201, 210, 212

appeals process, v, vii, viii, 1, 2, 3, 6, 7, 8, 11, 12, 14, 15, 17, 18, 21, 22, 23, 24, 31, 33, 40, 41, 42, 44, 45, 47, 48, 123

awareness, 195, 204, 206

B

basic needs, ix, 54, 156, 166

benefits, x, 56, 73, 122, 123, 126, 138, 139, 140, 143, 144, 145, 160, 161, 162, 182, 193

businesses, 131, 137, 138, 142, 143

C

call centers, 153, 177, 178, 211

challenges, vii, viii, x, xi, 2, 3, 7, 9, 15, 20, 27, 29, 33, 34, 38, 41, 45, 47, 48, 64, 65, 68, 70, 87, 93, 140, 148, 149, 151, 152, 166, 167, 168, 169, 171, 174, 175, 176, 177, 179, 182, 187, 189, 190, 191, 192, 196, 202, 203, 209, 210, 213, 214

children, 29, 55, 129

chronic illness, 156

civil rights, 151, 155, 183, 186, 192, 210, 213, 214, 215

communication systems, 149, 166

community, 30, 54, 55, 56, 58, 60, 69, 73, 74, 75, 83, 84, 85, 88, 91, 116, 126, 158, 163, 171, 172

coordination, 55, 125, 157, 160, 161, 164, 189, 195

cost, 10, 11, 57, 92, 124, 125, 127, 134, 159, 205

cost of living, 134

critical infrastructure, 157

D

damages, iv, ix, 5, 36, 58, 64, 66, 67, 74, 75, 83, 84, 85, 87, 89, 90, 128, 133, 141
data collection, 47, 92
data set, 21, 187
database, 211
Department of Health and Human Services, 134, 136, 141
Department of Homeland Security, 1, 3, 5, 6, 31, 42, 48, 49, 63, 65, 67, 134, 147, 151, 153, 173, 191, 210, 213, 216
Department of Labor, 127
disability, viii, xi, 129, 134, 141, 143, 148, 149, 150, 151, 152, 153, 154, 155, 156, 159, 160, 161, 162, 163, 164, 165, 171, 177, 179, 180, 181, 182, 183, 185, 186, 187, 188, 189, 190, 191, 192, 193, 194, 196, 197, 198, 199, 200, 201, 202, 203, 204, 205, 206, 209,210, 211, 212, 213, 214, 215
disability integration, viii, xi, 147, 148, 149, 150, 151, 152, 153, 154, 155, 160, 161, 162, 163, 164, 165, 187, 188, 192, 193, 194, 195, 196, 197, 198, 199, 200, 201, 202, 203, 204, 206, 209, 212, 213
disaster, vii, viii, ix, x, xi, 5, 9, 10, 22, 29, 36, 37, 38, 41, 53, 54, 55, 56, 57, 58, 59, 60, 64, 65, 67, 68, 69, 70, 71, 73, 74, 75, 78, 81, 83, 84, 85, 88, 89, 91, 92, 93, 95, 97, 121, 122, 123, 124, 126, 127, 128, 129, 130, 131, 132, 133, 134, 135, 137, 138, 139, 140, 141, 142, 143, 144, 145, 148, 150, 151, 152, 153, 155, 156, 157, 158, 159, 160, 161, 162, 164, 165, 168, 170, 171, 172, 173, 174, 176, 179, 181, 182, 183, 185, 187, 188, 190, 191, 192, 196, 198, 200, 203, 204, 205, 209, 210, 211, 213, 214, 215
disaster area, ix, 54, 57, 58, 124, 127, 131

disaster assistance, ix, xi, 5, 54, 58, 67, 71, 75, 123, 130, 133, 137, 138, 139, 140, 141, 143, 144, 148, 151, 152, 153, 157, 158, 159, 170, 172, 173, 181, 183, 185, 188, 190, 203, 205, 209, 211, 213, 214
Disaster Loan Program, viii, x, 121, 123, 134
disaster relief, 158
Disaster Relief Fund, 57
discrimination, 155, 160, 162, 183, 186, 192, 213, 214, 215

E

education, 164, 200
educational services, 56, 73, 126
emergency, x, 5, 8, 28, 29, 30, 32, 43, 47, 48, 55, 58, 64, 67, 69, 70, 71, 74, 75, 78, 79, 86, 89, 90, 91, 125, 144, 151, 152, 157, 158, 160, 161, 162, 163, 164, 165, 171, 173, 193, 194, 195, 197, 198, 199, 201, 202, 203, 210
emergency management, x, 8, 28, 29, 30, 32, 43, 47, 48, 64, 70, 78, 86, 89, 90, 91, 151, 157, 160, 161, 162, 163, 164, 171, 193, 194, 195, 197, 198, 199, 201, 202, 203
emergency planning, 201
emergency preparedness, 160, 210
employment, 56, 126, 132
equipment, 129, 157, 162, 166, 180, 182, 185, 201

F

families, 29, 55, 61, 67, 134, 182
federal agency, 144, 160
federal assistance, ix, x, 10, 29, 53, 58, 69, 71, 74, 78, 122, 129, 130, 137, 158, 161, 162

Index

223

Federal Emergency Management Agency, viii, ix, x, 1, 3, 4, 5, 8, 12, 22, 23, 24, 25, 27, 28, 38, 43, 45, 48, 53, 54, 63, 66, 72, 77, 78, 80, 81, 85, 89, 92, 97, 98, 109, 114, 115, 116, 117, 118, 121, 122, 129, 132, 134, 135, 137, 140, 141, 147, 149, 150, 151, 164, 165, 173, 178, 184, 195, 198, 209, 214, 215

federal government, viii, x, 5, 10, 60, 67, 71, 121, 140, 145, 152, 156, 157, 160

federal law, xi, 11, 29, 129, 148, 212

Federal Register, 91, 134, 140, 173

FEMA IA, x, 57, 79, 122, 123, 125, 129, 131, 133, 134

financial, viii, ix, x, 5, 53, 56, 57, 67, 73, 121, 127, 128, 129, 135, 138, 140, 143, 144, 151, 157, 158, 159, 160, 162, 173, 182

financial resources, 151

funding, ix, 5, 10, 20, 29, 30, 53, 54, 67, 158, 199, 205

funds, 5, 10, 23, 24, 29, 67, 82, 87, 97, 125, 132, 137, 140, 158

G

governments, 5, 10, 11, 29, 151, 156, 158, 170

governor, 9, 55, 61, 67, 71, 91, 125, 130, 158

grant programs, 5

grants, vii, viii, 2, 5, 22, 36, 71, 123, 130, 133, 158

guidance, 15, 37, 68, 69, 70, 90, 91, 94, 139, 143, 154, 161, 175, 176, 191, 196, 203, 205, 212

guidelines, 16, 134, 136

H

Hawaii, 80, 97, 100

health, 67, 151, 156, 157, 166, 167, 168, 170

Health and Human Services, 134

health condition, 156

hearing impairment, 169

hiring, 3, 29, 34, 35, 38, 41, 42, 43

home ownership, 69, 84, 85

homeowners, 124, 131, 137

homes, 67, 69, 74, 75, 76, 83, 84, 85, 89, 131, 167, 170, 180, 181

household income, 69, 132, 134, 142, 144

housing, 56, 57, 73, 76, 127, 136, 144, 159, 173, 176, 194

human, 38, 129, 138, 152, 197, 210

human capital, 38, 197

Hurricane Katrina, 156

Hurricane survivors, viii, xi, 148

hurricanes, xi, 5, 10, 148, 149, 151, 152, 154, 155, 156, 166, 168, 170, 176, 177, 181, 186, 191, 192, 210, 211, 212, 213

I

IA declaration requests, vii, x, 64, 65, 68, 76, 77, 79, 82, 84, 93

income, 56, 58, 60, 69, 74, 84, 88, 90, 92, 116, 126, 132, 133, 134, 135, 141, 142, 143, 145, 177

independent living, 186

Individual Assistance (IA) Program, v, vii, ix, 53, 54, 55, 123, 125, 213

individuals, vii, viii, ix, x, xi, 5, 34, 54, 56, 58, 61, 64, 65, 66, 71, 73, 75, 91, 121, 122, 123, 126, 128, 130, 137, 138, 143, 144, 148, 149, 150, 151, 152, 154, 155, 158, 159, 160, 161, 162, 163, 164, 165, 166, 167, 168, 170, 171, 172, 173, 174, 176, 177, 179, 180, 181, 182, 183, 185, 186, 187, 188, 189, 190, 191, 192, 193, 198, 199, 200, 201, 202, 203, 204, 209, 210, 211, 212, 213, 215

Index

individuals with disabilities, xi, 138, 148, 149, 150, 151, 154, 155, 160, 161, 162, 164, 165, 171, 173, 179, 180, 183, 185, 186, 187, 188, 191, 192, 198, 199, 201, 202, 204, 211, 212, 214

information sharing, 174

information technology, 160, 179

infrastructure, ix, 4, 53, 54, 60, 67, 91, 130, 180

integration, viii, xi, 148, 149, 150, 151, 152, 153, 154, 155, 160, 161, 162, 163, 164, 165, 187, 188, 192, 193, 194, 196, 197, 198, 199, 200, 201, 202, 203, 204, 206, 209, 212, 213

internal controls, 49, 86

issues, 3, 6, 14, 15, 17, 26, 27, 29, 31, 32, 33, 41, 43, 48, 49, 131, 154, 165, 179, 180, 182, 194, 213

L

large-scale disasters, 151

laws, 35, 68, 70, 151, 155, 161, 162, 183, 186, 192, 213, 214, 215

leadership, 94, 150, 158, 165, 174, 187, 194, 195, 198, 199

loans, 69, 122, 123, 124, 125, 130, 131, 132, 133, 142, 145

local authorities, ix, 54

local community, 168

local government, vii, viii, x, 2, 10, 29, 71, 78, 121, 124, 130, 137, 139, 140, 151, 156, 157, 158, 160, 161, 166, 171, 172, 173

M

major disaster, vii, ix, x, 5, 9, 53, 54, 55, 58, 60, 67, 68, 72, 73, 88, 91, 121, 122, 127, 128, 129, 130, 131, 137, 138, 144, 158, 161, 162, 178

major disaster declaration, vii, ix, 53, 54, 55, 60, 67, 68, 91, 122, 129, 130, 131

management, vii, viii, 2, 3, 6, 8, 13, 15, 16, 18, 21, 28, 30, 31, 33, 35, 37, 38, 40, 41, 45, 47, 48, 49, 70, 87, 89, 90, 151, 156, 163, 175, 176, 193, 194, 195, 196, 197, 200, 201, 202, 203, 205

media, 32, 129, 132, 134, 135, 137, 138, 169, 188

medical, vii, ix, 57, 64, 67, 129, 138, 157, 159, 162, 166, 167, 168, 170, 182, 185, 215

medical care, 166

medical expenses, vii, ix, 64, 182

medication, 129, 166, 167

N

natural disaster, 138

nonprofit organizations, xi, 5, 10, 131, 148, 158, 166, 175

O

officials, viii, x, xi, 2, 3, 7, 9, 10, 11, 13, 15, 17, 18, 19, 20, 21, 27, 28, 29, 30, 32, 33, 34, 35, 36, 37, 38, 39, 40, 43, 47, 48, 49, 64, 65, 68, 69, 70, 73, 74, 75, 78, 79, 82, 83, 84, 85, 87, 89, 90, 91, 93, 94, 148, 149, 151, 152, 153, 154, 155, 163, 164, 165, 166, 167, 168, 169, 170, 171, 172, 173, 174, 175, 176, 177, 179, 180, 181, 182, 183, 186, 187, 188, 189, 190, 191, 193, 194, 195, 196, 197, 198, 199, 200, 201, 202, 205, 210, 211, 212, 213, 214

organize, 14

outreach, 56, 73, 126, 162, 164, 187, 190, 196, 206

Index

225

P

PA appeals, vii, viii, 2, 6, 7, 11, 13, 14, 15, 17, 31, 33, 35, 36, 38, 39, 41, 42, 46, 47, 48

PA grant application, vii, viii, 2, 10

PA grants, vii, viii, 2, 36, 37

policy, xi, 11, 13, 15, 30, 31, 34, 35, 36, 37, 48, 122, 123, 134, 139, 142, 143, 144, 148, 154, 174, 183, 191, 212

positive relationship, 65, 86

potential benefits, 3, 34

poverty, 69, 84, 134, 136

preparedness, 152, 157, 163, 164, 166, 194, 204

press conferences, 169

Privacy Protection Act, 139

program staff, 14, 15, 175, 199

progress reports, 56, 127

project, 6, 10, 11, 15, 20, 28, 30, 32, 36, 37, 49

public health, 67, 152, 158, 210

Puerto Rico, xi, 67, 80, 81, 95, 97, 105, 108, 109, 133, 148, 149, 150, 151, 152, 166, 167, 168, 170, 171, 172, 178, 179, 180, 181, 182, 186, 187, 198, 201, 202, 210, 215

R

recommendations, iv, 2, 15, 16, 42, 43, 60, 61, 65, 84, 149, 161, 199, 201, 203, 205

recovery, viii, x, xi, 4, 9, 15, 36, 37, 56, 71, 75, 122, 126, 127, 129, 133, 135, 137, 148, 152, 157, 158, 160, 165, 172, 176, 198, 204

recovery assistance, viii, xi, 148, 173

recovery plan, 56, 127

recovery process, 56, 126

regulations, 6, 30, 58, 60, 68, 70, 130, 134, 139, 143, 212

repair, ix, 5, 30, 53, 56, 57, 124, 125, 127, 128, 130, 134, 142, 159, 182

resources, 4, 10, 32, 36, 37, 56, 57, 61, 67, 71, 90, 92, 126, 127, 137, 139, 157, 158, 159, 167, 183, 205

response, 4, 7, 9, 10, 14, 18, 32, 35, 37, 41, 43, 45, 59, 60, 67, 71, 129, 130, 137, 152, 153, 154, 157, 158, 160, 161, 165, 166, 172, 186, 198, 204, 210, 211, 212

S

safe shelter, viii, xi, 148

SBA Disaster Loan Program, x, 122, 123, 129, 130, 131, 133, 134

self-assessment, 154, 213

services, iv, 55, 56, 57, 58, 69, 73, 84, 88, 116, 125, 126, 127, 136, 151, 152, 160, 161, 168, 169, 170, 171, 183, 186, 188, 192, 193, 203, 210

shelter, vii, viii, ix, xi, 29, 64, 67, 148, 151, 162, 168, 181, 202

signs, 5

Small Business Administration (SBA), v, viii, x, xi, 69, 121, 122, 123, 124, 125, 128, 129, 130, 131, 132, 133, 134, 135, 136, 137, 139, 140, 141, 142, 143, 145, 146

social security, 132, 134, 141, 143

standardization, 19, 205

survivors, viii, x, xi, 56, 58, 73, 121, 124, 126, 140, 141, 148, 150, 151, 153, 155, 157, 161, 162, 166, 167, 177, 182, 187, 189, 191, 198, 200, 202, 203, 204, 210, 213

T

technical assistance, 158

technical comments, 43, 94, 205

temporary housing, ix, 29, 54, 57, 127, 138, 159, 162
territorial, xi, 5, 10, 57, 136, 148, 151, 158, 166, 167, 168, 169, 170, 201
territory, ix, 64, 152, 174, 186, 210
time frame, 6, 7, 8, 15, 16, 17, 26, 27, 28, 29, 32, 34, 39, 45, 46, 47
training, 3, 10, 14, 20, 31, 35, 36, 37, 38, 41, 42, 43, 69, 149, 150, 161, 164, 165, 188, 194, 197, 198, 199, 200, 201, 202, 203, 204, 205, 206
transportation, 29, 57, 128, 151, 159, 166, 167, 169, 171, 176, 180, 202
transportation infrastructure, 166, 202
trauma, 58, 67, 69, 74, 83, 84, 85

U.S. disasters, vii, viii, 2, 4
unemployment insurance, 73
United States, v, 1, 5, 44, 57, 63, 66, 67, 80, 81, 95, 97, 108, 109, 127, 138, 144, 147, 159, 182, 186, 207

victims, ix, 53, 56, 58, 59, 73, 75, 126
voluntary organizations, x, 121, 124, 156, 157, 158

workforce, 3, 32, 33, 34, 38, 41, 42, 43, 152, 153, 206, 212

Related Nova Publications

HURRICANES AND WILDFIRES: IMPACT, ASSISTANCE AND RECOVERY

EDITOR: Teri Boyd

SERIES: Natural Disaster Research, Prediction and Mitigation

BOOK DESCRIPTION: In 2017, four sequential disasters—hurricanes Harvey, Irma, Maria, and the California wildfires—created an unprecedented demand for federal disaster response and recovery resources. According to FEMA, 2017 included three of the top five costliest hurricanes on record.

HARDCOVER ISBN: 978-1-53614-895-4
RETAIL PRICE: $230

DISASTER SOCIAL WORK FROM CRISIS RESPONSE TO BUILDING RESILIENCE

AUTHOR: HC Johnston Wong

SERIES: Natural Disaster Research, Prediction and Mitigation

BOOK DESCRIPTION: This book describes the processes of crisis intervention, community mental health promotion and post traumatic growth. Putting resilience at heart has led to the ACT-R approach which can be learned not just by social workers, but by all response workers.

SOFTCOVER ISBN: 978-1-53614-435-2
RETAIL PRICE: $82

To see a complete list of Nova publications, please visit our website at www.novapublishers.com

Related Nova Publications

Emergency Management: An Overview and Issues for Congress

Editor: Stephanie Padbury

Series: Natural Disaster Research, Prediction and Mitigation

Book Description: After a flood, people are often uncertain if their eligibility for federal disaster assistance is linked to any way to whether or not they have flood insurance. The first two chapters in this book provide an overview of the assistance available to individuals and households following a flood and provides links to more comprehensive guidance on both flood insurance and disaster assistance.

Softcover ISBN: 978-1-53614-101-6
Retail Price: $82

Weathering the Storm: A State and Local Perspective on Emergency Management

Author: Anthony Rigg

Series: Natural Disaster Research, Prediction and Mitigation

Book Description: This book is the hearing that took place in front of the U.S. House of Representatives on June 10th 2011. It provides a State and Local Perspective on Emergency Management, specifically on the efforts of State, local and non-government organizations to prepare for and respond to natural disaster, terrorist attacks and other emergencies.

Softcover ISBN: 978-1-53613-788-0
Retail Price: $79

To see a complete list of Nova publications, please visit our website at www.novapublishers.com